D0122976

# PUBLISHERS' NOTE

The series of monographs in which this title appears was introduced by the publishers in 1957, under the General Editorship of Dr Maurice G. Kendall. Since that date, more than twenty volumes have been issued, and in 1966 the Editorship passed to Alan Stuart, D.Sc.(Econ.), Professor of Statistics, University of London.

The Series fills the need for a form of publication at moderate cost which will make accessible to a group of readers specialized studies in statistics or courses on particular statistical topics. Often, a monograph on some newly developed field would be very useful, but the subject has not reached the stage where a comprehensive treatment is possible. Considerable attention has been given to the problem of producing these books speedily and economically.

It is intended that in future the Series will include works on applications of statistics in special fields of interest, as well as theoretical studies. The publishers will be interested in approaches from any authors who have work of importance suitable for the Series.

CHARLES GRIFFIN & CO. LTD

# OTHER GRIFFIN BOOKS ON STATISTICS AND MATHEMATICS

*A statistical primer* F. N. DAVID
*A course in theoretical statistics* N. A. RAHMAN
*An introduction to the theory of statistics* G. U. YULE and M. G. KENDALL
*The advanced theory of statistics* M. G. KENDALL and A. STUART
*Rank correlation methods* M. G. KENDALL
*Exercises in theoretical statistics* M. G. KENDALL
*Exercises in probability and statistics (for mathematics undergraduates)* N. A. RAHMAN
*Rapid statistical calculations* M. H. QUENOUILLE
*Statistical communication and detection: with special reference to digital data processing of radar and seismic signals* E. A. ROBINSON
*Biomathematics* C. A. B. SMITH
*The design and analysis of experiment* M. H. QUENOUILLE
*Mathematical model building in economics and industry* C-E-I-R LTD
*Computer simulation models* JOHN SMITH
*Random processes and the growth of firms* J. STEINDL
*Sampling methods for censuses and surveys* F. YATES
*Statistical method in biological assay* D. J. FINNEY
*The mathematical theory of epidemics* N. T. J. BAILEY
*Probability and the weighing of evidence* I. J. GOOD
*Games, gods and gambling: the origins and history of probability and statistical ideas* F. N. DAVID
*Combinatorial chance* F. N. DAVID and D. E. BARTON

## GRIFFIN'S STATISTICAL MONOGRAPHS AND COURSES:

No. 1: *The analysis of multiple time-series* M. H. QUENOUILLE
No. 2: *A course in multivariate analysis* M. G. KENDALL
No. 3: *The fundamentals of statistical reasoning* M. H. QUENOUILLE
No. 4: *Basic ideas of scientific sampling* A. STUART
No. 5: *Characteristic functions* E. LUKACS
No. 6: *An introduction to infinitely many variates* E. A. ROBINSON
No. 7: *Mathematical methods in the theory of queueing* A. Y. KHINTCHINE
No. 8: *A course in the geometry of* n *dimensions* M. G. KENDALL
No. 9: *Random wavelets and cybernetic systems* E. A. ROBINSON
No. 10: *Geometrical probability* M. G. KENDALL and P. A. P. MORAN
No. 11: *An introduction to symbolic programming* P. WEGNER
No. 12: *The method of paired comparisons* H. A. DAVID
No. 13: *Statistical assessment of the life characteristic: a bibliographic guide* W. R. BUCKLAND
No. 14: *Applications of characteristic functions* E. LUKACS and R. G. LAHA
No. 15: *Elements of linear programming with economic applications* R. C. GEARY and M. D. MCCARTHY
No. 16: *Inequalities on distribution functions* H. J. GODWIN
No. 17: *Green's function methods in probability theory* J. KEILSON
No. 18: *The analysis of variance* A. HUITSON
No. 19: *The linear hypothesis: a general theory* G. A. F. SEBER
No. 20: *Econometric techniques and problems* C. E. V. LESER
No. 21: *Stochastically dependent equations: an introductory text for econometricians* P. R. FISK
No. 22: *Patterns and configurations in finite spaces* S. VAJDA
No. 23: *The mathematics of experimental design: incomplete block designs and Latin squares* S. VAJDA
No. 24: *Cumulative sum tests: theory and practice* C. S. VAN DOBBEN DE BRUYN

# MATHEMATICAL METHODS IN THE THEORY OF QUEUEING

A. Y. KHINTCHINE

*Translated by* D. M. ANDREWS, B.A. *and*
M. H. QUENOUILLE, M.A., Sc.D., F.R.S.E.

*Second Edition, with additional notes by*
ERIC WOLMAN, Ph.D.

---

BEING NUMBER SEVEN OF
GRIFFIN'S STATISTICAL
MONOGRAPHS & COURSES
EDITED BY
ALAN STUART, D.Sc.(Econ.)

---



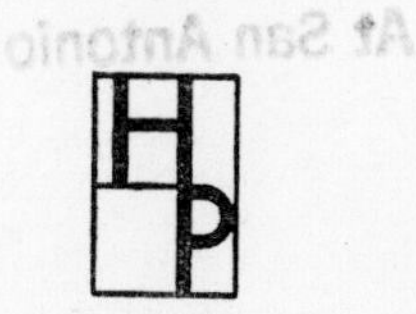

1969
HAFNER PUBLISHING COMPANY
NEW YORK

CHARLES GRIFFIN & COMPANY LIMITED
42 DRURY LANE, LONDON, WC2


Translated from the Russian text first published under the title *Matematicheskie metody teorii massovogo obsluzhivaniya* ("Mathematical methods of the theory of mass service"), as Vol. 49 (1955) of the series Trudy Matematicheskovo Instituta Steklov, Akad. Nauk, U.S.S.R.

---

*First published ... 1960*
*Second edition ... 1969*

PRINTED IN GREAT BRITAIN BY CHORLEY & PICKERSGILL LTD LEEDS

# PREFACE TO FIRST EDITION

In the realm of practical human affairs, situations frequently arise in which there is a demand for service in the face of which the serving organisation, since it provides only a limited number of serving units, is unable to satisfy all incoming demands. We all know examples of such situations. Queues in shops or at ticket offices, in refreshment bars, at hairdressers, etc.; the impossibility of getting a ticket for a train that one wishes to travel on, owing to its being booked up; the delay in landing an aircraft caused by the lack of free runways; the delay in repairing a machine which has broken down, on account of the lack of repair gangs—all these and many other examples like them are, apart from the substantial divergence of their actual content, very similar to one another from the point of view of form. In all such instances our theory has one main task—to establish as clearly as possible the interdependence of the number of serving units and the quality of the service. As regards this, the quality of the service in different instances is, naturally, measured by different indices. Usually, either the percentage of demands which are turned down (the percentage of passengers that did not get a ticket for a given train) or the average time of waiting for the commencement of the service (queues of different kinds) is taken as the index. Obviously, to achieve a higher quality of service requires in all cases a greater number of service units. However, it is evident that an excessive number will result in a superfluous outlay of effort and material resources. Thus, in practice, the problem is usually resolved by the quality of the service initially finding its necessary level, and thereafter by discovering the minimal number of service units by which this level can be attained.

In problems of a similar kind it is nearly always necessary to take into account the influence of the element of chance upon the course of the phenomenon under consideration. The quantity of incoming demands is not as a rule constant, but undergoes occasional fluctuation. The time of serving the demands is not in the majority of problems a standard one, but is subject to fluctuation from one problem to another.

All these chance elements do not by any means have the character of little "disturbances", breaking up the even and regular course of the phenomenon; on the contrary they form the main feature in our picture of the processes to be studied. Thus it is natural that the concepts and methods of the theory of probability—a mathematical discipline devoted to the study of the frequencies of events—should become a mathematical instrument of the theory of queueing.

The aim of the present work is to acquaint the reader with the main ideas, methods, and different ways of thought which govern the application of the theory of probability to questions of mass service. The need for a monograph of this kind has long been felt both by mathematicians and by practical workers (in the first instance by those who, in their own time, brought the theory under consideration into being and have remained up till now its chief users). This need is so much the more reinforced by the fact that few available explanations of the general theory exist, even outside Russia.

In itself, of course, the present monograph cannot in any way pretend to present all the information there is to be communicated. The theory of queueing has developed considerably in recent times, and for only a fractional explanation, or even to put down the most important of its achievements, a large volume would be required. However, I have set myself no such task. My object has been to throw some light on the main character of the most important examples, i.e. on the method of employing probability arguments in questions of mass service, and from this "methodological" statement of aims I have drawn the choice of material for my monograph. Special attention has been paid to the chief exponent of the theory, A. K. Erlang, whose investigations are not by any means known to the degree they deserve. I have also incorporated into the book a significant number of the ideas of C. Palm, who is doing most at present to carry on Erlang's work. To a certain extent I have also joined in the attempt—one which continues in our own time—to find simpler methods of investigation in contrast with the analytical methods which have been in favour almost without a break from the time of Erlang's classic work, and which required the establishment of differential, partial-differential, and integral equations. As far as this is concerned I would draw the reader's attention to §§ 2 and

25 and also to the whole of Chapter 11. An attempt to form a general elementary theory was undertaken a short while ago in an interesting article by Lundkvist (*Ericsson Technics*, 1953).

The preparation of this monograph was greatly hampered by the fact that all the fundamental literature comes from the pens of workers in the practical field, and is therefore unsatisfactory from a mathematical point of view. To give each explanation a form that was mathematically acceptable I was not able to leave a single discussion in its original state; it was necessary either substantially to supplement the author's reasoning or to reject it altogether and substitute a different argument. Similarly, where new concepts were being introduced it was necessary in many cases to define them differently, since the definitions given by the authors seemed insufficiently precise.

In order to help the reader to link the general concepts with concrete phenomena, I use terminology borrowed from telephonic practice during the course of the whole book — hitherto questions in this field have constituted the greatest stimulus to the development of the theory of mass service. Thus I speak of "calls" (instead of "demands" or "requirements"), of "losses" (instead of "refusals"), of the length of a "conversation" (instead of a "service") etc. However, in all instances everything that is said can be naturally related to any kind of mass service, by making the necessary changes in terminology.

The book is divided into three parts of which the first, the largest, is devoted to the study of the stream of incoming calls and contains no questions of service at all (the other two parts being devoted to this). It seemed appropriate to confine the study of the incoming stream to a particular part of the book, not only because, in order to provide a good service, it is necessary to know what you are serving, but also because the theory of an incoming stream of calls is none other than the general theory of the random sequence of uniform occurrences, finding a broad circle of applications even in fields unconnected with any kind of service (for example, in questions of the radioactive disintegration of atoms).

I have tried to make the book accessible to anyone who has mastered the main concepts of the theory of probability and followed some course, however short, in mathematical analysis. In many instances the explanation could have been abbreviated if, instead of introducing the full proof of this or that assertion, I had allowed myself to appeal to the results of an appropriate theory of probability (Markoff chains, random processes, ergodic theories). However, wishing to produce a book accessible to as wide as possible a circle of readers, I have hardly ever yielded to such a temptation.

In the theory of probability, it is usual to denote the distribution function of a given random variable $\xi$ by a non-diminishing function $F(x)$, which gives the probability of the inequality $\xi < x$. In the theory of mass service it is often more convenient to express the distribution function of $\xi$ by a non-increasing function $\Phi(x)$, which gives the probability of the inequality $\xi > x$. Depending on circumstances, we will use either of these expressions for the distribution function.

A. Y. KHINTCHINE

## EDITORIAL NOTE TO SECOND EDITION

For this new edition the translation of Khintchine's well-known text appears substantially unchanged, apart from correction of a few translation errors and minor misprints. However, some improvements have been made, which take the form of additional notes amplifying or amending the theory in certain details; these are printed at the end of the book, with footnote references in the text. Our grateful thanks are due to Dr Eric Wolman of Bell Telephone Laboratories, U.S.A., for supplying these notes. We also acknowledge the kind assistance of Mr A. Jhavary of the Statistics Research Division, London School of Economics, who has overhauled the Bibliography and corrected some errors in the Russian original.

October 1968 A.S.

# CONTENTS

CONTENTS

# *PART ONE*

# THE INCOMING STREAM OF CALLS

## CHAPTER 1

## THEORY OF THE SIMPLE STREAM

The general theory of streams (of uniform events) might reasonably begin with abstract definitions of the main components of such streams. However, any such general approach will be postponed till Chapter 2 where it will be dealt with to the extent necessary. Meanwhile, we prefer to introduce the reader into the field of concrete investigations of streams of the simplest type in order to achieve from the start a clear picture of the basic ways of thought, of the mathematical models in the theory of mass service, and of the form of this science as a mathematical discipline. When these concrete explanations have been sufficiently mastered, the study of a more abstract general theory should not seem difficult.

We also add that the simple type of stream to be studied in this chapter has, in course of time, shown itself to be almost the only one useful in application; only in comparatively recent years has the necessity been clearly shown for the study of streams of a more general type. Meanwhile, even nowadays the great majority of applications of the theory of mass service (in particular in telephone applications) arise from the assumption that an incoming stream of demands (calls) belongs to this simple type. Explanations, particularly technical ones, of the theory of the simple stream have in the past few decades developed to the extent that even elementary courses in the theory of probability nowadays usually include a special chapter devoted to this theory.

### 1 Definition and statement of the problem

We will call a stream of uniform events "simple" if it possesses the following three characteristics —

(1) *Stationarity*. For any $t > 0$ and integer $k \geqslant 0$, the probability that during the period of time $(a, a + t)$ there will occur $k$ events is the same for every $a \geqslant 0$ (and so depends only on $t$ and $k$); hereafter we shall indicate this probability by $v_k(t)$. Throughout the book we shall deal only with streams in which, during a finite period of time with probability 1, there occur

only a finite number of events. Thus, for any $t$, we shall always have

$$\sum_{k=0}^{\infty} v_k(t) = 1.$$

The stationarity of a stream expresses the invariability of its probability regime in time.

(2) *Absence of after-effects.* The probability $v_k(t)$ of the occurrence of $k$ events during the period $(a, a + t)$ does not depend on the sequence of events up to the moment $a$; in other words, the conditional probability of the occurrence of $k$ events during the period $(a, a + t)$ calculated for any supposition about the sequence of events up to the moment $a$ is equal to the unconditional probability $v_k(t)$ of that event. The absence of after-effects expresses the mutual independence of sections of a stream in periods of time which do not overlap.

(3) *Orderliness.* Suppose that for a given stationary stream $\psi(t)$ signifies the probability that during a period of time of length $t$ there occur at least two events.

(Thus $\psi(t) = 1 - v_0(t) - v_1(t) = \sum_{k=2}^{\infty} v_k(t)$). Then we have

$$\psi(t) = o(t) \qquad (t \to 0)$$

or, equally,

$$\frac{\psi(t)}{t} \to 0, \qquad (t \to 0).$$

As will be seen later, the orderliness of a stream expresses the practical impossibility of two or more events occurring in the same moment of time.

*Thus, we call a "simple stream of uniform events" any stationary, orderly stream without after-effects.*

The main task in the theory of the simple stream is the definition of the type of functions $v_k(t)$; putting it in a different way, our objective will be to find the distribution function of the number of events during a period of time of length $t$, regarded as a random variable. To provide a real interpretation of the stream under consideration, it will be assumed in what follows that the stream of calls is coming into some telephone exchange, and correspondingly our uniform events will be termed "calls".

## 2 Elementary solution

Let us divide the period of time (0,1) into an arbitrary number $n$ of equal periods of length $1/n$. The probability that in any one of these divisions there will occur no calls at all is equal to $v_0(1/n)$; and as our stream is without after-effects,

$$v_0(1) = \left[ v_0\left(\frac{1}{n}\right) \right]^n,$$

or supposing $v_0(1) = \theta$,

$$v_0\left(\frac{1}{n}\right) = \theta^{1/n}.$$

If we have a period of length $k/n$ ($k = 1, 2, \ldots$), then it can be broken down into $k$ periods of length $\frac{1}{n}$, in consequence of which

$$v_0\left(\frac{k}{n}\right) = \left[ v_0\left(\frac{1}{n}\right) \right]^k = \theta^{k/n}. \tag{2.1}$$

Finally, suppose that $t$ is any positive number and that the integer $k$ is determined by the inequality

$$\frac{k-1}{n} < t \leqslant \frac{k}{n};$$

then since $v_0(t)$ is a non-increasing function of $t$,

$$v_0\left(\frac{k-1}{n}\right) \geqslant v_0(t) \geqslant v_0\left(\frac{k}{n}\right)$$

or, by virtue of (2.1),

$$\theta^{\frac{k-1}{n}} \geqslant v_0(t) \geqslant \theta^{k/n}.$$

But, as $n \to \infty$, we have $k/n \to t$, in consequence of which the extreme terms of the above inequalities tend to $\theta^t$, and we find

$$v_0(t) = \theta^t$$

for any $t > 0$. In this context, the parameter $\theta$ is defined as $v_0(1)$, and consequently, $0 \leqslant \theta \leqslant 1$. The cases $\theta = 0$, and $\theta = 1$, however, do not interest us, and we need not concern ourselves with them.

Actually, when $\theta = 1$ we have $v_0(t) = 1$ for all $t > 0$, which signifies a complete absence of calls in any period of time, i.e. the absence of any stream whatever. Further, if $\theta = 0$, then $v_0(t) = 0$ for any $t > 0$; this means that actual calls will be received in any period of time at all, however small; but then, no matter how great were $k$, the number of calls (authentic) within any period would be greater than $k$; in other words the number of calls in any period is infinite with probability 1; but in Section 1 such streams were excluded from consideration once and for all. Thus it may be assumed that $0 < \theta < 1$; and hence that $\theta = e^{-\lambda}$, where $\lambda$ is a constant positive number, so that

$$v_0(t) = e^{-\lambda t}. \tag{2.2}$$

In this deduction the orderliness of our stream has been nowhere used, so that the relation (2.2) is valid for any stationary stream without after-effects; this observation will be important later on.

Now let us consider the functions $v_k(t)$ when $k > 0$. There are many different ways of solving this problem, and nearly all of them instructive, since methods founded upon them help to solve a whole series of more complicated problems. We shall begin with the most elementary method. We shall regard $t$ as a constant, and divide the period $(0, t)$ into an arbitrary number $n > k$ of equal parts (compartments) of length $t/n = \delta$. According to the distribution of calls within these compartments, two outcomes are possible —

$H_1$ — there will not be more than one call in any one of the $n$ compartments.

$H_2$ — there will be more than one call in at least one of the compartments.

Then it follows that

$$v_k(t) = P(H_1,k) + P(H_2,k) \tag{2.3}$$

where $P(H_i,k)$ $(i = 1, 2)$ signifies the probability of a double event — (1) outcome $H_i$ is realised, and (2) in the period $(0, t)$ there occur $k$ calls. Evidently $P(H_1,k)$ is the probability of any state of affairs in which, out of the $n$ compartments, $k$ contain one call and the rest $n-k$ do not contain any calls. Thus

$$P(H_1,k) = \binom{n}{k}\left[v_1(\delta)\right]^k\left[v_0(\delta)\right]^{n-k}.$$

By virtue of formula (2.2) and the orderliness of the given stream we have, as $n \to \infty$ $(\delta \to 0)$ and for constant $k$,

$$[v_0(\delta)]^{n-k} = e^{-\lambda\delta(n-k)} = e^{-\lambda t}\, e^{k\lambda\delta} = e^{-\lambda t}\,[1 + o(1)]$$

$$[v_1(\delta)]^k = [1-e^{-\lambda\delta}-\psi(\delta)]^k = [1-e^{-\lambda\delta}+o(\delta)]^k$$

$$= (\lambda\delta)^k\left[1 + o(1)\right] = \frac{(\lambda t)^k}{n^k}\left[1 + o(1)\right]$$

and consequently

$$P(H_1,k) = e^{-\lambda t}\,\frac{(\lambda t)^k}{k!}\;\frac{n(n-1)\ldots(n-k+1)}{n^k}\left[1 + o(1)\right]$$

$$\to e^{-\lambda t}\,\frac{(\lambda t)^k}{k!} \qquad (n \to \infty)$$

On the other hand, $P(H_2,k)$ does not exceed the probability of $H_2$, i.e. that at least one of the $n$ compartments will contain more than one call. Just as the probability of one of the individual compartments containing more than one call is $\psi(\delta)$, so

$$P(H_2,k) \leqslant n\,\psi(\delta) = t\,\frac{\psi(\delta)}{\delta} \to 0 \qquad (n \to \infty)$$

In this way the right-hand side of the equation (2.3) as $n \to \infty$ has the limit

$$e^{-\lambda t} \frac{(\lambda t)^k}{k!}$$

and since the left-hand side of (2.3) does not depend on $n$,

$$v_k(t) = e^{-\lambda t} \frac{(\lambda t)^k}{k!} \qquad (k = 0,1 \ldots)$$

Thus for a simple stream the number of calls in a period of length $t$ is distributed in a Poisson distribution with parameter $\lambda t$.

## 3 Method of differential equations

In specialist literature the problem to be solved is usually approached by another, less elementary, method, but one which, for this reason, is easily extended to cover more complicated problems. This method will now be considered.

Suppose $t$ and $\tau$ to be any positive numbers and suppose that $k > 0$. Let $k_1$ be the number of calls in the period $(0, t)$ and $k_2$ in the period $(t, t + \tau)$. For $k$ calls to occur in the period $(0, t + \tau)$ it is necessary and sufficient for one of the following double events to happen — $k_1 = k$, $k_2 = 0$; $k_1 = k-1$, $k_2 = 1$; $k_1 = k - 2$, $k_2 = 2$; . . .; $k_1 = 0$, $k_2 = k$. But the probabilities that $k_1 = l$ and $k_2 = m$ are respectively equal to $v_l(t)$ and $v_m(\tau)$; and as these instances are mutually independent (a process without after-effects!) it follows that

$$v_k(t+\tau) = v_k(t)v_0(\tau) + v_{k-1}(t)v_1(\tau) + v_{k-2}(t)v_2(\tau) + \ldots + v_0(t)v_k(\tau) \qquad (3.1)$$

But, as $\tau \to 0$, we have

$$v_0(\tau) = e^{-\lambda\tau} = 1 - \lambda\tau + o(\tau)$$

$$v_1(\tau) = 1 - v_0(\tau) - \psi(\tau) = \lambda\tau + o(\tau)$$

$$\sum_{l=2}^{k} v_{k-l}(t)v_l(\tau) \leqslant \sum_{l=2}^{\infty} v_l(\tau) = \psi(\tau) + o(\tau)$$

and consequently (3.1) gives

$$v_k(t+\tau) = v_k(t)\,(1-\lambda\tau) + v_{k-1}(t)\lambda\tau + o(\tau)$$

$$\frac{v_k(t+\tau) - v_k(t)}{\tau} = \lambda[v_{k-1}(t) - v_k(t)] + o(1).$$

This shows that the function $v_k(t)$ is differentiable for any $t > 0$ and that

$$v_k'(t) = \lambda[v_{k-1}(t) - v_k(t)] \qquad (k = 1,2, \ldots) \qquad (3.2)$$

If for convenience it is supposed that $v_{-1}(t) \equiv 0$, then the equation (3.2) is true even when $k = 0$, as may be confirmed directly from (2.2).

Thus, we have derived a system of linear differential equations for the required functions $v_k(t)$. This system is easily solved by several methods of which we shall now consider the two most instructive.

(*A*) *Method of transforming the required functions*

Supposing

$$v_k(t) = e^{-\lambda t}u_k(t) \qquad (k = 0,1,2, \ldots)$$

hence

$$v_k'(t) = e^{-\lambda t}[u_k'(t) - \lambda u_k(t)] \qquad (k = 0,1,2, \ldots)$$

and inserting these expressions for $v_k(t)$ and $v_k'(t)$ into equation (3.2) it follows that

$$u_k'(t) = \lambda u_{k-1}(t) \qquad (k = 0,1,2, \ldots)$$

where by definition $u_{-1}(t) \equiv 0$. Hence, by integration

$$u_k(t) - u_k(0) = \lambda \int_0^t u_{k-1}(z)\, dz \qquad (k = 0,1,2 \ldots).$$

Evidently, for any $k \geqslant 0$, we have

$$v_k(0) = u_k(0)$$

but, by definition of the functions $v_k(t)$,

$$v_0(0) = 1,\ v_k(0) = 0 \qquad (k > 0).$$

Thus $u_0(0) = 1$, $u_k(0) = 0$ $(k = 1,2, \ldots)$ and it follows that, when $k \geqslant 1$,

$$u_k(t) = \lambda \int_0^t u_{k-1}(z)\, dz. \tag{3.3}$$

Observing that, by virtue of (2.2) and from the definition of the functions $u_k(t)$,

$$u_0(t) \equiv 1,$$

and using formula (3.3) recurrently we find that

$$u_1(t) = \lambda t$$

$$u_2(t) = \frac{(\lambda t)^2}{2}$$

$$\cdots\cdots$$

$$u_k(t) = \frac{(\lambda t)^k}{k!}.$$

Consequently

$$v_k(t) = e^{-\lambda t}\frac{(\lambda t)^k}{k!}$$

i.e. the previous solution to the problem is obtained

(*B*) *Method of generating functions*

Suppose that

$$\sum_{k=0} v_k(t)x^k = \Phi\,(t, x),$$

then the series on the left-hand side of this equation converges absolutely when $|x| \leqslant 1$. By multiplying all terms of the equation (3.2) by $x^k$ and summing for $k$ from 0 to $\infty$ we find

$$\frac{\partial \Phi}{\partial t} = \lambda \sum_{k=0}^{\infty} v_{k-1}(t)x^k - \lambda\Phi = \lambda(x-1)\Phi,$$

or

$$\frac{\partial \log_e \Phi}{\partial t} = \lambda(x-1).$$

Hence

$$\log_e \Phi(t, x) - \log_e \Phi(0, x) = \lambda(x-1)t. \tag{3.4}$$

But it is easy to see that for any $x$

$$\Phi(0, x) = v_0(0) = 1.$$

Thus (3.4) gives

$$\Phi(t,x) = e^{\lambda(x-1)t} = e^{-\lambda t}e^{\lambda tx} = e^{-\lambda t}\sum_{k=0}^{\infty}\frac{(\lambda t)^k}{k!}x^k.$$

Comparing this with the definition of function $\Phi(t, x)$ it may be seen that

$$v_k(t) = e^{-\lambda t}\frac{(\lambda t)^k}{k!} \qquad (k = 0,1,2,\ldots)$$

i.e. the same solution is again derived.

## 4 Intensity of a simple stream

The above results show that the three properties (stationarity, absence of after-effects, orderliness) by which we defined a simple stream, fully characterise its structure apart from the value of the parameter $\lambda$, which may take any positive value. Two simple streams can differ from one another only according to the value of this parameter.

Henceforth, we will denote by $w(t)$ the probability that for any stationary stream there will occur at least one call during a length of time $t$. Then we have

$$w(t) = 1 - v_0(t) = \sum_{k=1}^{\infty} v_k(t) = v_1(t) + \psi(t)$$

where $\psi(t) = \sum_{k=2}^{\infty} v_k(t)$ signifies, as before, the probability of at least two calls during a period of time of length $t$. For a simple stream with parameter $\lambda$, $v_0(t) = e^{-\lambda t}$ and, as $t \to 0$,

$$w(t) = 1 - e^{-\lambda t} = \lambda t + o(t)$$

or

$$\lim_{t\to 0}\frac{w(t)}{t} = \lambda. \tag{4.1}$$

This may be considered as a relation *defining* the parameter $\lambda$ for a given stream. Later, it will be shown that the limit (4.1) exists for any stationary stream and that the parameter $\lambda$, defined by the relation (4.1), is one of the most important characteristics of this stream.

But we shall return now to the simple stream and find the mathematical expectation of the number of calls coming in over a period of length $t$. This is equal to

$$\sum_{k=1}^{\infty} kv_k(t) = e^{-\lambda t} \sum_{k=1}^{\infty} \frac{(\lambda t)^k}{(k-1)!} = e^{-\lambda t}\, \lambda t \sum_{k=1}^{\infty} \frac{(\lambda t)^{k-1}}{(k-1)!} = \lambda t$$

since the last summation is equal to $e^{\lambda t}$. This result might have been foreseen earlier, as we know that the mathematical expectation of a variable distributed according to Poisson's law is equal to the parameter of this law, i.e. here equals $\lambda t$.

The mathematical expectation of the number of calls in a unit of time is called the "intensity" of a given stream; this intensity will be denoted by $\mu$. As has just been established, $\mu = \lambda$ for a simple stream. However, for stationary streams of more complex structure this equation is not only not evident, but is not even always true. This question will be examined in detail further on. For the moment, it will be sufficient to establish that for any stationary stream $\mu \geqslant \lambda$.

Indeed, the mathematical expectation of the number of calls during time $t$ for a given stream is

$$\mu t = \sum_{k=1}^{\infty} kv_k(t) \geqslant \sum_{k=1}^{\infty} v_k(t) = w(t)$$

$$\mu \geqslant \frac{w(t)}{t}$$

and since the left-hand side of this inequality does not depend on $t$, by virtue of (4.1) $\mu \geqslant \lambda$.

*Thus, the intensity $\mu$ of a simple stream coincides with its parameter $\lambda$; at present, for any stationary stream it can only be stated that $\mu \geqslant \lambda$.* In this connection, the existence of the limit (4.1) for any stationary stream has still to be proved(*).

## 5 A stream with a variable parameter

In this book we shall usually confine our studies to stationary streams of calls. However, for some of the simplest problems, the solution in non-stationary situations is so easy to deduce and also of such clear practical value that it would be a pity to leave it aside without examining

---

(*) In essence the parameter $\lambda$ itself can be interpreted as the "intensity" of a given stream, since the relation (4.1) shows that the probability of the occur rence of calls in an infinitesimally small period of length $t$ is asymptotically proportional to $t$, and the parameter $\lambda$ is the coefficient of proportion. It would be possible to speak of a "higher intensity" and a "lower intensity" (since $\mu \geqslant \lambda$ always). Later in § 11 it will be shown that for an orderly stream $\mu = \lambda$ always.

it at all. In this section in particular, we shall study non-stationary streams which are, like simple streams, orderly and without after-effects. Let us explain these proposals.

If a stream is not stationary, the probability of receiving $k$ calls in a period of length $\tau$ depends not only upon $\tau$, but also on the first moment $t$ of this period — therefore it will be designated by $v_k(\tau,t)$. Thus $v_k(\tau,t)$ is the probability that during the period $(t,t+\tau)$ there occur $k$ calls. By analogy with the stationary instance we write

$$1 - v_0(\tau,t) = w(\tau,t), \quad 1 - v_0(\tau,t) - v_1(\tau,t) = \psi(\tau,t).$$

The resultant stream will be called orderly if as $\tau \to 0$, for any constant $t \geqslant 0$, the following relation holds —

$$\frac{\psi(\tau,t)}{\tau} \to 0.$$

Furthermore, it is necessary that, for any $t \geqslant 0$, there exists

$$\lim_{\tau \to 0} \frac{w(\tau,t)}{\tau} = \lambda(t) \tag{5.1}$$

(the instantaneous value of the parameter).

With these assumptions, it is now necessary to find an expression for the functions $v_k(\tau,t)$. First, consider as before, the case $k = 0$.

Since we are dealing with a stream without after-effects, for $\Delta\tau > 0$

$$v_0(\tau+\Delta\tau,t) = v_0(\tau,t)v_0(\Delta\tau,t+\tau).$$

But according to the above assumption, as $\Delta\tau \to 0$ and when $t$ and $\tau$ are constants,

$$v_0(\Delta\tau,t+\tau) = 1 - w(\Delta\tau,t+\tau) = 1 - \lambda(t+\tau)\Delta\tau + o(\Delta\tau).$$

Consequently

$$v_0(\tau+\Delta\tau,t) - v_0(\tau,t) = - v_0(\tau,t)\lambda(t+\tau)\Delta\tau + o(\Delta\tau).$$

Dividing both sides by $\Delta\tau$ gives in the limit the relation

$$\frac{\partial v_0(\tau,t)}{\partial\tau} = - \lambda(t+\tau)v_0(\tau,t), \tag{5.2}$$

(whereby the existence of the derivative is incidentally shown). Hence

$$\frac{\partial \log_e v_0(\tau,t)}{\partial\tau} = - \lambda(t+\tau),$$

and consequently

$$\log_e v_0(\tau,t) - \log_e v_0(0,t) = - \int_0^\tau \lambda(t+u)du.$$

Since $\log_e v_0(0,t) = 0$, it follows that

$$\log_e v_0(\tau,t) = -\int_0^\tau \lambda(t+u)du = -\Lambda(\tau,t)$$

$$v_0(\tau,t) = \exp - \Lambda(\tau,t).$$

In the stationary case, the index was $-\lambda\tau$. In the general case, as we shall now see, $\lambda$ should be replaced by the magnitude

$$\frac{1}{\tau}\Lambda(\tau,t) = \frac{1}{\tau}\int_0^\tau \lambda(t+u)du$$

which it is natural to regard as the mean value of the "instantaneous parameter" $\lambda(t)$ in the period $(t, t+\tau)$.

Consider now the case $k > 0$. By analogy with the above (*vide* § 3) it follows that as $\Delta\tau \to 0$ and if $t, \tau$ are constants,

$$v_k(\tau+\Delta\tau,t) = v_k(\tau,t)\, v_0(\Delta\tau,t+\tau) + v_{k-1}(\tau,t)\, v_1(\Delta\tau,t+\tau) + o(\Delta\tau)$$

where

$$v_0(\Delta\tau,t+\tau) = 1 - \lambda(t+\tau)\,\Delta\tau + o(\Delta\tau)$$

and

$$v_1(\Delta\tau,t+\tau) = 1 - v_0(\Delta\tau,t+\tau) - \psi(\Delta\tau,t+\tau) = \lambda(t+\tau)\Delta\tau + o(\Delta\tau);$$

consequently,

$$v_k(\tau+\Delta\tau,t) = v_k(\tau,t)\,[1 - \lambda(t+\tau)\Delta\tau] + v_{k-1}(\tau,t)\lambda(t+\tau)\Delta\tau + o(\Delta\tau).$$

From which

$$\frac{v(\tau+\Delta\tau,t) - v_k(\tau,t)}{\Delta\tau} = \lambda(t+\tau)\,[v_{k-1}(\tau,t) - v_k(\tau,t)] + o(1),$$

and hence, in the limit,

$$\frac{\partial v_k(\tau,t)}{\partial\tau} = \lambda(t+\tau)\,[v_{k-1}(\tau,t) - v_k(\tau,t)]. \tag{5.3}$$

This relation, which has been established for any $k > 0$, remains true, as (5.2) shows, even for $k = 0$ if we set

$$v_{-1}(\tau,t) \equiv 0.$$

The necessary solution of system (5.3) will now be found by employing the method of generating functions. Suppose that

$$F(t,\tau,x) = \sum_{k=0}^{\infty} v_k(\tau,t)x^k.$$

Multiplying all terms of the equation (5.3) by $x^k$ and summing for $k$ from 0 to $\infty$, we find, by an exact analogy with § 3,

$$\frac{\partial F}{\partial \tau} = (x-1)\lambda(t+\tau)F,$$

or

$$\frac{\partial \log_e F}{\partial \tau} = (x-1)\lambda(t+\tau),$$

whence

$$\log_e F(t,\tau,x) - \log_e F(t,0,x)$$

$$= (x-1)\int_0^\tau \lambda(t+u)du = (x-1)\Lambda(\tau,t). \tag{5.4}$$

For any $x$ and $t$,

$$F(t,0,x) = v_0(0,t) = 1.$$

Thus (5.4) gives

$$F(t,\tau,x) = e^{(x-1)\Lambda(\tau,t)} = e^{-\Lambda(\tau,t)}\, e^{x\Lambda(\tau,t)}$$

$$= \sum_{k=0}^{\infty} e^{-\Lambda(\tau,t)} \frac{[\Lambda(\tau,t)]^k}{k!} x^k,$$

and a comparison with the definition of function $F(t,\tau,x)$ gives

$$v_k(\tau,t) = e^{-\Lambda(\tau,t)} \frac{[\Lambda(\tau,t)]^k}{k!} \qquad (k = 0,1,2\ldots). \tag{5.5}$$

These formulae completely solve the problem in hand. It may be seen that even for a stream with a variable parameter the number of calls in the period $(t,t+\tau)$ is subject to Poisson's law; however, the parameter of this law now depends not only on the length $\tau$ of the given period but also on the first moment $t$. For a stationary stream, Poisson's law with parameter $\lambda\tau$ held; in the transition to a non-stationary case, the constant $\lambda$ must be replaced by the expression

$$\frac{\Lambda(\tau,t)}{\tau} = \frac{1}{\tau}\int_0^\tau \lambda(t+u)du,$$

i.e. the mean value of $\lambda(x)$ in the period $t \leqslant x \leqslant t+\tau$. In particular, if the function $\lambda(x) = \lambda$ is a constant magnitude, then, obviously, for any $t$

$$\Lambda(\tau,t) = \int_0^\tau \lambda(t+u)du = \lambda\tau,$$

and the formulae (5.5) become the solutions derived in § 3 for the stationary case.

Finally, it should be observed that the number of calls in the period $(t,t+\tau)$, subject to Poisson's law, has as its mathematical expectation the parameter of this law, i.e. the magnitude

$$\Lambda(\tau,t) = \int_t^{t+\tau} \lambda(u)du.$$

Thus one may take the magnitude $\frac{1}{\tau}\Lambda(\tau,t)$ as the *mean* intensity of our stream in period $(t,t+\tau)$; and the limit of this magnitude, as $\tau \to 0$, is the *instantaneous* intensity $\mu(t)$ of the given stream at the moment $t$. It thus follows that

$$\mu(t) = \lim_{\tau \to 0} \frac{1}{\tau}\Lambda(\tau,t) = \lim_{\tau \to 0} \frac{1}{\tau}\int_t^{t+\tau} \lambda(u)du = \lambda(t).$$

Thus, even for a simple stream with a variable parameter, the instantaneous intensity of the stream equals the instantaneous value of the parameter.

# CHAPTER 2

# GENERAL PROPERTIES OF STATIONARY STREAMS

## 6 The stream of demands as a random process

As was said at the beginning of Chapter 1, the streams of demands which are encountered can in most instances be reasonably approximated by simple streams. Thus, in studying such streams, the results given in Chapter 1 will generally be used. However, in recent years, with science being set ever more complex problems and being required to give ever more accurate solutions, the author has found it necessary to study also streams of a more general nature. There are two immediate causes for this widening of the field of study. On the one hand, observations on streams of demands, even in the most ordinary circumstances, are showing increasingly that deductions based on the assumption of a simple stream do not correspond sufficiently well with the experimental data (it is easily possible to foresee such discrepancies theoretically — it is not difficult to show, for example, that in practice it is usually advisable to expect a certain type of after-effect in a given stream, and that this cannot always be neglected). On the other hand, there are cases when the stream in question differs from a simple stream obviously and fundamentally. All kinds of streams of variable intensity are of this type (they are not stationary, *vide* § 5); so are streams entering in the second, third, etc. line of a "fully accessible collection", even when a simple stream enters into the first line; all these streams have a significant after-effect increasing with the number of the line, and acknowledgment of this after-effect is necessary in the theory (*vide* Chapter 8). Therefore present-day investigations into the theory of mass service cannot confine themselves to the investigation of simple streams and must extend their initial assumptions.

In turning to the investigation of streams of a more general type, an exact definition of the main concepts is necessary in the interests of strictness and unambiguous clarity; this was not provided in Chapter 1, as that chapter served as an introductory one whose purpose was to demonstrate simply the types of mathematical methods applicable to the whole theory of streams.

If $x(t)$ denotes the number of calls coming in during the period $(0,t)$, then, for each fixed value $t > 0$, $x(t)$ represents a random variable. With variable $t$, $x(t)$ represents a one-parameter family of random variables, which is called a "random process" or "random function". It is characteristic of the function $x(t)$ that: (1) it can take only non-negative integer values, and (2) with increasing $t$, it never decreases.

Thus a graph of such a function, regardless of any individual case, always has the form of a series of steps, as depicted in Fig. 1.

In order for any random process $x(t)$ to be specified, it is necessary that for any finite group of positive numbers $t_1, t_2, \ldots, t_n$ there should be given an $n$-dimensional distribution function of the vector

$$x(t_1),\ x(t_2),\ \ldots,\ x(t_n).$$

If the process $x(t)$ represents a stream of calls and, in consequence, $x(t)$ can take only non-negative integer values, then in order to describe this stream as a random process it is necessary to know for each group of positive numbers $t_1, t_2, \ldots, t_n$ and for each group of non-negative integers $k_1, k_2, \ldots, k_n$ the probability of the system of equations $x(t_1) = k_1,\ x(t_2) = k_2, \ldots, x(t_n) = k_n$ (obviously this probability can differ from zero only if, when $t_1 < t_2 < \ldots < t_n$, we also have $k_1 \leqslant k_2 \leqslant \ldots \leqslant k_n$). In particular ($n = 1$), for any $t > 0$ and any non-negative integer $k$, the probability of the equation $x(t) = k$ which was

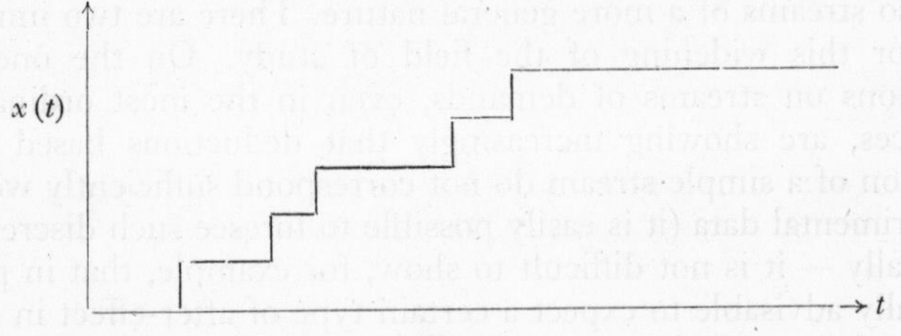


**Fig. 1**

denoted by $v_k(t)$(*) in Chapter 1 must be known. Thus the system of functions $v_k(t)$ ($k = 0,1,2\ldots$) is required in the composition of a description of each stream of calls. In the majority of cases, however, the description of this system of functions is still insufficient fully to characterise the stream.

A stream of calls is termed "stationary" if, for $0 < t_1 < t_2 < \ldots < t_n$ and for any positive $a$, the distribution function of the vector $x(t_i)$ $(1 \leqslant i \leqslant n)$ coincides with the distribution function of the vector $x(a + t_i) - x(a)$ $(1 \leqslant i \leqslant n)$; in other words the distribution function of the vector $x(a + t_i) - x(a)$ $(1 \leqslant i \leqslant n)$ depends on the figure $t_i$, but does not depend on $a$. In particular when $n = 1$, $v_k(t)$ for a stationary stream signifies the probability of the occurrence of $k$ calls in the period $(a, a + t)$, where $a \geqslant 0$ arbitrarily (i.e. in any period of length $t$)(†).

(*) One should observe, however, that in Chapter 1 $v_k(t)$ (by virtue of the proposed stationarity of the stream) indicated the probability of $k$ incoming calls in *any* period of time of length $t$, whereas now we are only concerned with the period $(0,t)$.

(†) For readers acquainted with the theory of random processes, we observe that a stationary stream of calls defined in this way is not, of course, a stationary random process in the generally accepted sense of that term. In the theory of random processes our stationary stream belongs to the group of "streams with stationary increments".

A given stream of calls is termed a "stream without after-effects" if the distribution function of the vector $x(a + t_i) - x(a)$ $(t_i > 0, 1 \leqslant i \leqslant n)$ for any $a \geqslant 0$ does not depend on the value of the magnitude $x(t)$ for any values of $t < a$. This definition, of course, expresses in an exact form the requirement that the chance sequence of a stream of calls after any moment of time "$a$" should be independent of its sequence up to the moment "$a$"; but this also implies the absence of after-effects within the sense of the theory of probability.

It is easy to see that a stationary stream without after-effects is fully characterised by the system of functions $v_k(t)$ $(k = 0,1,2 \ldots)$, i.e. by the distribution function (where there is a suitable one) of the number of calls entering in the course of a period of time of length $t$. Indeed, since the system of equations $x(t_i) = k_i$ $(1 \leqslant i \leqslant n, 0 < t_1 < t_2 < \ldots < t_n)$ is obviously equivalent to the system of equations

$$x(t_i) - x(t_{i-1}) = k_i - k_{i-1} \qquad (1 \leqslant i \leqslant n),$$

where in general the proposed $t_0 = k_0 = x(t_0) = 0$, then (if $P\{\ \}$ denotes the probability of the event located within the curly brackets we shall have

$$P\{x(t_i) = k_i, \quad 1 \leqslant i \leqslant n\} = P\{x(t_i) - x(t_{i-1}) = k_i - k_{i-1}, \quad 1 \leqslant i \leqslant n\}.$$

As the periods $(t_0,t_1)$, $(t_1,t_2)$, $\ldots$, $(t_{n-1},t_n)$ do not overlap, we obtain from this, in the absence of after-effects,

$$P\{x(t_i) = k_i, 1 \leqslant i \leqslant n\} = \prod_{i=1}^{n} P\{x(t_i) - x(t_{i-1}) = k_i - k_{i-1}\},$$

and since, by virtue of the stationarity of the stream,

$$P\{x(t_i) - x(t_{i-1}) = k_i - k_{i-1}\} = P\{x(t_i - t_{i-1}) = k_i - k_{i-1}\} = v_{k_i - k_{i-1}}(t_i - t_{i-1}) \quad (1 \leqslant i \leqslant n),$$

it follows that[*]

$$P\{x(t_i) = k_i, 1 \leqslant i \leqslant n\} = \prod_{i=1}^{n} v_{k_i - k_{i-1}}(t_i - t_{i-1}).$$

This shows that the set of functions $v_k(t)$ uniquely determines the probabilities $P\{x(t_i) = k_i, 1 \leqslant i \leqslant n\}$, i.e. the given stream is fully characterised as a random process.

In particular, for a simple stream with parameter $\lambda$ we obtained in § 2

$$v_k(t) = e^{-\lambda t} \frac{(\lambda t)^k}{k!} \qquad (k = 0,1,2 \ldots).$$

---

[*] Here and hereafter we shall put $v_r(t) = 0$ for any negative index $r$.

Thus for a simple stream with parameter $\lambda$, it follows that when $0 = t_0 < t_1 < \ldots < t_n$, $0 = k_0 \leqslant k_1 \leqslant \ldots \leqslant k_n$

$$P\{x(t_i) = k_i, 1 \leqslant i \leqslant n\}$$
$$= \prod_{i=1}^{n} e^{-\lambda(t_i - t_{i-1})} \frac{\lambda^{k_i - k_{i-1}}(t_i - t_{i-1})^{k_i - k_{i-1}}}{(k_i - k_{i-1})!}$$
$$= e^{-\lambda t_n} \lambda^{k_n} \prod_{i=1}^{n} \frac{(t_i - t_{i-1})^{k_i - k_{i-1}}}{(k_i - k_{i-1})!}$$

## 7 Fundamental properties of stationary streams

As in Chapter 1, $w(t)$ will be used to denote the probability that for any stationary stream there will occur at least one call in a period of time of length $t$, so that

$$w(t) = 1 - v_0(t) = \sum_{k=1}^{\infty} v_k(t).$$

In § 4 it was established that for a stationary stream without after-effects (and in particular for a simple stream) the expression $w(t)/t$ tends to a fixed limit $\lambda$ as $t \to 0$. The parameter of the given stream was denoted by $\lambda$, and this plays an important part in the study of the basic properties of this stream. The relation

$$\frac{w(t)}{t} \to \lambda \qquad (t \to 0) \tag{7.1}$$

is equivalent to the relation

$$w(t) = \lambda t + o(t) \qquad (t \to 0) \tag{7.2}$$

and is often expressed by saying that, for small $t$, $w(t)$ is asymptotically proportional to $t$. In the great majority of expositions of the theory of the simple stream the relation (7.2) or (7.1) is included directly in the definition of a simple stream(*) and this, as we have seen, is unnecessary, since this relation emerges as a consequence of the requirements of stationarity and absence of after-effects.

However, the relation (7.1) has a significantly wider field of applicability — it is true, as will be shown, for *any stationary stream.* Thus every stationary stream, regardless of the presence or absence of after-effects, has a parameter as defined in Chapter 1. This circumstance, as will be seen, is most convenient for the study of the general properties of stationary streams.

The proof of the existence of the limit (7.1) for any stationary stream is based on the following elementary lemma of limit theory, which will be of use to us later on.

(*) *Vide* Erlang [7], Feller [3], Fry [4], Khintchine [5].

*Lemma.* Suppose the function $f(x)$ is non-negative and non-decreasing in the segment $0 < x \leqslant a$, and $f(x + y) \leqslant f(x) + f(y)$ if $x,y$ and $x + y$ belong to the segment $(0,a)$. Then as $x \to 0$ the ratio $f(x)/x$ either increases to infinity or else tends to some limit; this limit is equal to zero only in the trivial instance of $f(a) = 0$.

*Proof.* It follows naturally from the inequality $f(x + y) \leqslant f(x) + f(y)$ that

$$f(x) \leqslant m f\left(\frac{x}{m}\right) \tag{7.3}$$

for $0 < x \leqslant a$ and for any integer $m$. In particular, when $x = a$,

$$\frac{f\left(\frac{a}{m}\right)}{\frac{a}{m}} \geqslant \frac{f(a)}{a} \qquad (m = 1,2\ldots).$$

This shows that (excluding the trivial instance $f(a) = 0$)

$$\alpha = \overline{\lim_{x\to 0}} \frac{f(x)}{x} \geqslant \frac{f(a)}{a} > 0$$

in which case $\alpha = +\infty$ is not excluded.

Let us assume at the start that $\alpha < +\infty$. Let the number $c > 0$ be such that

$$\frac{f(c)}{c} > \alpha - \varepsilon \tag{7.4}$$

where $\varepsilon > 0$ is arbitrarily small, and let $0 < x < c$. Determine the integer $m \geqslant 2$ from the inequality

$$\frac{c}{m} \leqslant x < \frac{c}{m - 1}.$$

Then by virtue of (7.3) and the assumed monotonicity of $f(x)$,

$$\frac{f(x)}{x} \geqslant \frac{f\left(\frac{c}{m}\right)}{\frac{c}{m-1}} = \frac{m-1}{m}\,\frac{mf\left(\frac{c}{m}\right)}{c} \geqslant \frac{m-1}{m}\frac{f(c)}{c} \tag{7.5}$$

and so, by virtue of (7.4)

$$\frac{f(x)}{x} \geqslant \left(1 - \frac{1}{m}\right)(\alpha - \varepsilon).$$

Since $\varepsilon$ is arbitrarily small and since $m \to \infty$ as $x \to 0$, it follows that $\lim_{x\to 0} \frac{f(x)}{x} = \alpha$ and the lemma is proved. The argument remains much the same for $\alpha = +\infty$. We take any large $A > 0$ and choose a number $c$ such that $f(c)/c > A$; then, from (7.5)

$$\frac{f(x)}{x} \geqslant \frac{m-1}{m} A,$$

and hence $f(x)/x \to +\infty$ as $x \to 0$.

*Theorem. For any stationary stream it is true that*

$$\lim_{t\to 0}\frac{w(t)}{t}=\lambda>0$$

*in which the instance $\lambda=+\infty$ is not excluded.*

*Proof.* It is sufficient to show that the function $w(t)$ in a given segment $(0,a)$ satisfies all the requirements of the lemma which has just been proved. It is evident that $w(t)\geqslant 0$ and with increasing $t$ cannot decrease; it is also apparent that $w(a)>0$ for sufficiently large $a$, if we leave aside the trivial instance of a stream in which calls are not possible at all. Finally, if there occurred at least one call in the period $(0,t_1+t_2)$, then, naturally, the same thing would occur for at least one of the two periods $(0,t_1)$ and $(t_1,t_1+t_2)$, whence

$$w(t_1+t_2)\leqslant w(t_1)+w(t_2)\ (t_1>0,\ t_2>0,\ t_1+t_2<a).$$

Thus this last hypothesis is fulfilled. Applying the lemma, the proof of the theorem follows directly.

If the given stationary stream is a stream without after-effects, then for it, as was shown in § 2, we have

$$v_0(t)=e^{-\lambda t}\qquad(\lambda>0\text{ constant}),$$

excluding cases where in any period of time there occur either an infinite number of calls or no calls at all; these two instances, as they have no practical significance, have been excluded from further investigation. Thus for a stream without after-effects

$$w(t)=1-e^{-\lambda t},\qquad \lim_{t\to 0}\frac{w(t)}{t}=\lambda.$$

The limit, the existence of which was proved in the last theorem, in the case of a stream without after-effects is always, therefore, some finite positive number. But if the possibility of after-effects is not excluded, then $\lambda$ may change to $+\infty$ even in situations which have not been excluded from our investigations.

In order to demonstrate this, consider the following example of a stationary stream. Imagine a simple stream, whose parameter $\lambda$ represents a random variable with distribution function $F(x)$ $[F(+0)=0,$ $F(+\infty)=1]$. At the end of § 6 we gave the probability of the system of equations $x(t_i)=k_i$ $(1\leqslant i\leqslant n)$ for the simple stream with a fixed value of the parameter $\lambda$; now, for the sake of brevity we will denote this probability by $P_\lambda(t_i,k_i)$. If the parameter $\lambda$ is a random variable with distribution function $F(x)$, then the probability of the system of equations $x(t_i)=k_i$ $(1\leqslant i\leqslant n)$ will (by the formula for compounding probability) be equal to

$$P(t_i,k_i)=\int_0^\infty P_\lambda(t_i,k_i)dF(\lambda). \tag{7.6}$$

Thus this probability is defined for any $n$, $t_i$, $k_i$ and, as we know, this completely describes the specified stream. This stream will clearly be stationary; but, generally speaking, it will be a stream with after-effects. If $w(t)$ and $v_k(t)$ have the usual meanings for the stream (7.6), and $w_\lambda(t)$ and $v_{k\lambda}(t)$ denote these magnitudes for a simple stream with parameter $\lambda$, then clearly,

$$w(t) = \int_0^\infty w_\lambda(t) dF(\lambda)$$

and

$$w_\lambda(t) = 1 - v_{0\lambda}(t) = 1 - e^{-\lambda t},$$

whence

$$\frac{w(t)}{t} = \int_0^\infty \frac{1 - e^{-\lambda t}}{t} dF(\lambda).$$

If

$$\int_0^\infty \lambda dF(\lambda) < +\infty$$

(i.e. the distribution $F(x)$ has a finite mathematical expectation), then since $1 - e^{-\lambda t} \leqslant \lambda t$ we get

$$\frac{w(t)}{t} \leqslant \int_0^\infty \lambda dF(\lambda).$$

The expression $w(t)/t$ as $t \to 0$ is thus limited, and the stream in question has a finite parameter. But, if the integral

$$\int_0^\infty \lambda dF(\lambda) \tag{7.7}$$

diverges, then $w(t)/t \to +\infty$ as $t \to 0$. In fact, if $\varepsilon > 0$ is arbitrarily small and $A$ is so big that

$$\int_0^A \lambda dF(\lambda) > \frac{1}{\varepsilon}$$

then, since

$$\frac{w(t)}{t} \geqslant \int_0^A \frac{1 - e^{-\lambda t}}{t} dF(\lambda) \to \int_0^A \lambda dF(\lambda) \qquad (t \to 0)$$

it follows that

$$\lim_{t \to 0} \frac{w(t)}{t} \geqslant \frac{1}{\varepsilon},$$

and hence, $w(t)/t \to +\infty$ $(t \to 0)$. Thus, when the integral (7.7) diverges, our stream has an infinite value for the parameter. Further, this stream gives with probability 1 a finite number of calls in any finite period of time. We have

$$v_k(t) = \int_0^\infty v_{k\lambda}(t)dF(\lambda),$$

or, since $v_{k\lambda}(t) = e^{-\lambda t}(\lambda t)^k/k!$

$$v_k(t) = \frac{t^k}{k!}\int_0^\infty \lambda^k e^{-\lambda t}dF(\lambda) > 0,$$

$$\sum_{k=0}^\infty v_k(t) = \int_0^\infty e^{-\lambda t}\sum_{k=0}^\infty \frac{(\lambda t)^k}{k!}dF(\lambda) = \int_0^\infty dF(\lambda) = 1.$$

## 8 General form of a stationary stream without after-effects

As was seen in Chapter 1, a stationary stream without after-effects, if it also has the property of orderliness is a simple stream, the general structure of which can easily be established. We now set ourselves the task of finding a general description of a stationary stream without after-effects, leaving aside the requirement of orderliness.

In § 6, it was seen that a stationary stream without after-effects is uniquely defined by its functions $v_k(t)$ ($k = 0,1,2\ldots$) (wherein $v_0(t)$ always equals $e^{-\lambda t}$). Thus our task is reduced to defining the general character of the family of functions $v_k(t)$ for stationary streams without after-effects. With this objective, it will first be established that for any such stream with $k > 0$ the expression $v_k(t)/t$ tends to a fixed limit (which may be either zero or a positive number) as $t \to 0$.

Denote by $\psi_k(t)$ ($k = 0,1,2\ldots$) the probability that there occur at least $k$ calls in a period of length $t$, so that for $k = 0,1,2\ldots$

$$\psi_k(t) = \sum_{i=k}^\infty v_i(t), \quad v_k(t) = \psi_k(t) - \psi_{k+1}(t),$$

$$\psi_0(t) = 1, \quad \psi_1(t) = w(t), \quad \psi_2(t) = \psi(t).$$

It was proved in § 7 that for any stationary stream the expression $\psi_1(t)/t$ as $t \to 0$ tends to a fixed limit, finite or infinite. For a stream without after-effects this result is trivial, since $\psi_1(t) = w(t) = 1 - e^{-\lambda t}$, from which $\psi_1(t)/t \to \lambda$ when $t \to 0$. But on the other hand we shall now see that for a stationary stream without after-effects the limit of the expression $\psi_k(t)/t$ as $t \to 0$ exists for any $k > 0$. Since $v_k(t) = \psi_k(t) - \psi_{k+1}(t)$, it follows that as $t \to 0$ even the limit of the expression $v_k(t)/t$ exists for any $k > 0$. In the case of a simple stream, of course, we get $v_1(t)/t \to \lambda$, $v_k(t)/t \to 0$ ($k > 1$).

By analogy with the argument in § 7, we shall begin by proving an elementary auxiliary proposition of limit theory which will supplement the lemma in § 7.

*Lemma. If the function $f(x)$ is non-negative and non-decreasing in the segment $0 < x \leqslant a$, if the expression $f(x)/x$ is bounded within this segment, and if*

$$f(nx) \leqslant nf(x) + cn^2x^2, \tag{8.1}$$

*where $c > 0$ is a constant, $n$ is any integer and $0 < nx \leqslant a$; then the expression $f(x)/x$ tends to a fixed limit $l \geqslant 0$ as $x \to 0$.*

*Proof.* Setting $x = x_0/n$ (where $0 < x_0 \leqslant a$) in (8.1), we get

$$f(x_0) \leqslant nf\left(\frac{x_0}{n}\right) + cx_0^2,$$

or

$$\frac{f\left(\frac{x_0}{n}\right)}{\frac{x_0}{n}} \geqslant \frac{f(x_0)}{x_0} - cx_0. \tag{8.2}$$

We may suppose that

$$\varliminf_{x\to 0} \frac{f(x)}{x} = l,$$

we may also take $l > 0$, since the proof of the lemma is trivial if $l = 0$.

Let $\varepsilon > 0$ be arbitrarily small. Let us choose $x_0 < \dfrac{\varepsilon}{c}$ such that

$$\frac{f(x_0)}{x_0} > l - \varepsilon.$$

If $0 < x < x_0$ and the integer $n > 1$ is defined by the inequalities

$$\frac{x_0}{n} \leqslant x < \frac{x_0}{n-1},$$

then, by virtue of (8.2)

$$\frac{f(x)}{x} \geqslant \frac{f\left(\frac{x_0}{n}\right)}{\frac{x_0}{n-1}} = \frac{n-1}{n}\,\frac{f\left(\frac{x_0}{n}\right)}{\frac{x_0}{n}} \geqslant \frac{n-1}{n}\frac{f(x_0)}{x_0} - cx_0$$

$$\geqslant \left(1 - \frac{1}{n}\right)(l - \varepsilon) - \varepsilon > l - 3\varepsilon,$$

if $x$ is sufficiently small. Since, on the other hand, $f(x)/x < l + \varepsilon$ for sufficiently small $x$, so

$$\frac{f(x)}{x} \to l \qquad (x \to 0).$$

and the lemma is proved.

To prove the existence of the limit of the expression $\psi_k(t)/t$ as $t \to 0$, it need only be shown that for $k > 0$ the function $\psi_k(t)$ satisfies all conditions of the proved lemma. Non-negativeness and monotonicity of $\psi_k(t)$ in any segment are self-evident. Further, from $\psi_k(t) \leqslant \psi_1(t) = w(t)$ and $w(t)/t \to \lambda$ $(t \to 0)$ it follows that $\psi_k(t)/t$ is bounded in any segment. Thus all there remains to do is to establish that $\psi_k(t)$ satisfies the inequality (8.1).

For this purpose, $g_l$ will be used to denote the maximum value of the expressions $\psi_l(t)/t$ in the range $0 < t < +\infty$ $(l \geqslant 1)$ and it will be supposed that

$$A_k = \sum_{l=1}^{k-1} g_l\, g_{k-l}.$$

It may be stated that

$$\psi_k(nt) \leqslant n\psi_k(t) + A_k \frac{n(n-1)}{2} t^2 \tag{8.3}$$

from which it will follow that the function $\psi_k(t)$ satisfies the inequality (8.1).

The inequality (8.3) will be proved by an induction on $n$. If $n = 1$ it is trivial. Assume that it holds for some $n$. For not less than $k$ calls to occur in the segment $[0,(n+1)t]$ of length $(n+1)t = nt + t$ (the probability of which is equal to $\psi_k\{(n+1)t\}$), it is essential that for any $l$ $(0 \leqslant l \leqslant k)$ there should not be less than $l$ calls in the segment $(0,t)$ and not less than $k - l$ calls in the segment $[t,(n+1)t]$ (of length $nt$). Therefore

$$\begin{aligned}\psi_k[(n+1)t] &\leqslant \sum_{l=0}^{k} \psi_l(t)\psi_{k-l}(nt)\\ &= \psi_0(t)\psi_k(nt) + \psi_k(t)\psi_0(nt) + \sum_{l=1}^{k-1} \psi_l(t)\psi_{k-l}(nt)\\ &\leqslant \psi_k(nt) + \psi_k(t) + nt^2 \sum_{l=1}^{k-1} g_l g_{k-l}\\ &= \psi_k(nt) + \psi_k(t) + A_k nt^2.\end{aligned}$$

By virtue of (8.3), it then follows that

$$\psi_k[(n+1)t] \leqslant (n+1)\psi_k(t) + A_k \frac{n(n+1)}{2} t^2$$

and the inequality (8.3) is true also for $n + 1$ as well as $n$.

Therefore the function $\psi_k(t)$ for $k > 0$ satisfies all the requirements of the proved lemma and consequently, as $t \to 0$, the ratio $\psi_k(t)/t$ and equally the ratio $v_k(t)/t$ tend to a fixed limit. Since $w(t)/t \to \lambda > 0$ as $t \to 0$, the ratio $v_k(t)/w(t)$ will also have a fixed limit.

Suppose that

$$\lim_{t\to 0} [v_k(t)/w(t)] = p_k \qquad (k = 1,2, \ldots).$$

The ratio $v_k(t)/w(t)$ is the probability of receiving $k$ calls in a segment of length $t$, *if it is certain that in this segment calls do exist.* Thus, as $t \to 0$ it is possible to regard the limit of this expression, i.e. the quantity $p_k$, as the probability of receiving $k$ calls at a chosen moment if it is certain that calls do occur generally at that moment. (Such an interpretation of the magnitude $p_k$ is possible, but, of course, not obligatory.)

Let us now consider the determination of the general class of functions $v_k(t)$ for a stationary stream without after-effects. It was established in § 3 that, for such a stream (*vide* (3.1)),

$$v_k(t+\tau) = \sum_{l=0}^{k} v_l(\tau)v_{k-l}(t) \qquad (t > 0,\ \tau > 0,\ k = 0,1,2\ldots).$$

Since, as $\tau \to 0$,

$$v_0(\tau) = e^{-\lambda\tau} = 1 - \lambda\tau + o(\tau),$$

it follows that when $k > 0$

$$v_k(t+\tau) = (1-\lambda\tau)v_k(t) + \sum_{l=1}^{k} v_l(\tau)v_{k-l}(t) + o(\tau)$$

and consequently that

$$\frac{v_k(t+\tau) - v_k(t)}{\tau} = -\lambda v_k(t) + \sum_{l=1}^{k} \frac{v_l(\tau)}{\tau} v_{k-l}(t) + o(1).$$

But, when $l > 0$ and $\tau \to 0$, it has been proved above that

$$\frac{v_l(\tau)}{\tau} = \frac{v_l(\tau)}{w(\tau)} \frac{w(\tau)}{\tau} \to \lambda p_l.$$

Thus the passage to the limit shows the existence of $v_k'(t)$ and gives

$$v_k'(t) = -\lambda v_k(t) + \lambda \sum_{l=1}^{k} p_l v_{k-l}(t) \qquad (k = 1,2,\ldots) \tag{8.4}$$

Adding the obvious relation

$$v_0'(t) = -\lambda v_0(t) \tag{8.5}$$

gives a system of equations which permits unique determination of the class of functions $v_k(t)$. In particular, the transformation of the unknown functions which was used in § 3, leads to this. Assuming, as there, that

$$v_k(t) = e^{-\lambda t}u_k(t) \qquad (k = 0,1,2,\ldots)$$

the system (8.4) is reduced to the form

$$u_k'(t) = \lambda[p_1u_{k-1}(t) + p_2u_{k-2}(t) + \ldots + p_ku_0(t)]$$

allowing the functions $u_k(t)$ (and also $v_k(t)$) to be determined by a recurrence formula. Thus, for example, by virtue of $u_0(t) = 1$ we find

$$u_1'(t) = \lambda p_1$$

whence

$$u_1(t) = \lambda p_1 t, \quad v_1(t) = e^{-\lambda t}\,\lambda p_1 t.$$

No further conclusions will be derived here, as much simpler and more pleasing results are given by the method of generating functions, the application of which will now be demonstrated.

Suppose that

$$F(t,x) = \sum_{k=0}^{\infty} v_k(t)x^k.$$

Then the search for the system of functions $v_k(t)$ is reduced to the search for the function $F(t,x)$. Multiplying the relation (8.4) (for $k = 0$ the alternative relation (8.5)) by $x^k$ and summing for $k$ from 0 to $\infty$, we find

$$\frac{\partial F}{\partial t} = - \lambda F + \lambda \sum_{k=1}^{\infty} x^k \sum_{l=1}^{k} p_l \, v_{k-l}(t)$$

$$= - \lambda F + \lambda \sum_{l=1}^{\infty} p_l \sum_{q=0}^{\infty} v_q(t)x^{q+l}$$

$$= - \lambda F + \lambda \sum_{l=1}^{\infty} p_l \, x^l \sum_{q=0}^{\infty} v_q(t)x^q$$

or setting

$$\sum_{l=1}^{\infty} p_l \, x^l = \Phi(x)$$

we get

$$\frac{\partial F}{\partial t} = \lambda[\Phi(x) - 1]F,$$

$$\frac{\partial \log_e F}{\partial t} = \lambda[\Phi(x) - 1].$$

Since, for any $x$

$$F(0,x) = v_0(0) = 1,$$

by integrating for $t$ it follows that

$$F(t,x) = \exp \lambda[\Phi(x) - 1]t \tag{8.6}$$

and the problem is solved. It should be noted that for any $t$

$$F(t,1) = \sum_{k=0}^{\infty} v_k(t) = 1$$

in consequence of which (8.6) gives

$$\Phi(1) = \sum_{l=1}^{\infty} p_l = 1.$$

Thus the generating function $F(t,x)$ for any stationary stream without after-effects satisfies (8.6) where $\lambda > 0$ and

$$\Phi(x) = \sum_{l=1}^{\infty} p_l \, x^l, \quad p_l \geqslant 0, \quad \sum_{l=1}^{\infty} p_l = 1.$$

Now let us establish, on the other hand, that, provided the parameters $\lambda$ and $p_l$ $(l = 1,2, \ldots)$ are subject to the aforementioned requirements, there exists a stationary stream without after-effects whose generating function is given by the formula (8.6).

For this purpose it will be assumed that moments of time at which calls occur follow a simple stream with parameter $\lambda$. Hence, for the probability $v_k^*(t)$ that during period $t$ there will occur $k$ of such "calling moments", we have the usual expression

$$v_k^*(t) = e^{-\lambda t} \frac{(\lambda t)^k}{k!}.$$

However, a stream of such calls will not usually be simple, since at each calling moment with a non-zero probability it is possible that more than one call may enter. Assume that the probability of the occurrence at a given calling moment of precisely $l$ calls equals $p_l$ $(l = 1,2, \ldots)$ independently of what kind of calling moment it is or what the course of this stream is up to the given moment. This will define a stream of calls which, of course, will be a stationary stream without after-effects. It will now be shown that the generating function $F(t,x)$ of this stream is obtained by formula (8.6).

The number of calls occurring at any calling moment has been defined as a random variable bearing the value $l$ with probability $p_l$ $(l = 1,2, \ldots)$. The generating function of this variable is

$$\sum_{l=1}^{\infty} p_l \, x^l = \Phi(x).$$

Now let us take $r$ of such calling moments, differing among themselves, and denote by $P_r(k)$ the probability that at these $r$ moments as a collection there will occur $k$ calls, so that the total number of such calls over $r$ calling moments is a random variable with generating function

$$\sum_{k=0}^{\infty} P_r(k) x^k$$

(where, of course, $P_r(k) = 0$ when $k < r$). But this random magnitude is the sum of $r$ mutually independent random magnitudes, each of which has the generating function $\Phi(x)$. Since in the addition of mutually independent random magnitudes their generating functions are multiplied together[*], it follows that

$$\sum_{k=0}^{\infty} P_r(k) x^k = \{\Phi(x)\}^r.$$

[*] This emerges directly from the fact that for the random variable with a distribution function $P\{\xi = n\} = q_n$ $(n = 0,1,2, \ldots)$ the generating function $f(x) = \sum_{n=0}^{\infty} q_n \, x^n$ is, clearly, the mathematical expectation of the variable $x^{\xi}$.

And since, on the other hand, it is evident that for the stream in question

$$v_k(t) = \sum_{r=0}^{\infty} v_r^*(t) P_r(k),$$

so

$$F(t,x) = \sum_{k=0}^{\infty} v_k(t) x^k = \sum_{k=0}^{\infty} x^k \sum_{r=0}^{\infty} v_r^*(t) P_r(k)$$

$$= \sum_{r=0}^{\infty} v_r^*(t) \sum_{k=0}^{\infty} P_r(k) x^k = \sum_{r=0}^{\infty} e^{-\lambda t} \frac{(\lambda t)^r}{r!} \{\Phi(x)\}^r$$

$$= e^{-\lambda t} \sum_{r=0}^{\infty} \frac{\{\lambda t\, \Phi(x)\}^r}{r!} = e^{\lambda t[\Phi(x)-1]}$$

which agrees with formula (8.6).

The result of this investigation may be stated as follows — the population of stationary streams without after-effects coincides with the population of all streams represented by the formula (8.6), where $\lambda > 0$ and

$$\Phi(x) = \sum_{l=1}^{\infty} p_l\, x^l, \quad p_l \geqslant 0\ (l = 1,2,\ldots), \quad \sum_{l=1}^{\infty} p_l = 1.$$

From an objective point of view, it has been established that for every stationary stream without after-effects the stream of calling moments is simple, and in order to describe fully the given call stream it is necessary not only to give the parameter $\lambda$ of this simple stream but also the distribution $(p_1, p_2, \ldots p_n, \ldots)$ of the number of calls entering at any selected calling moment. Obviously, these considerations clarify the structure of the general stationary stream without after-effects.

We note also that in the case $p_1 = 1$, $p_k = 0$ $(k > 1)$, formula (8.6) gives the generating function of a simple stream with parameter $\lambda$:

$$F(t,x) = e^{\lambda t(x-1)} = \sum_{k=0}^{\infty} \left\{ e^{-\lambda t} \frac{(\lambda t)^k}{k!} \right\} x^k.$$

CHAPTER 3

# PALM'S FUNCTIONS

## 9 Definition and proof of existence

Having discussed, in the last chapter, the structure of stationary streams without after-effects, we turn now to the investigation of streams of a more general type. For fairly broad types of such streams a very useful method of investigation is provided by a type of function introduced by Palm [8] and applied by him successfully to the solution of a series of problems. Palm defines this function $\varphi_0(t)$ (for any stationary stream) as *the conditional probability of the absence of calls in the period* $(t_0,t_0+t)$, *if it is certain that at the moment* $t_0$ *there occurred a call.* However, one can hardly regard such a definition as suitable — the condition under which the probability $\varphi_0(t)$ is to be calculated, i.e. the presence of a call at a certain moment $t_0$, has itself in all actual instances the probability 0, and this circumstance, naturally, prevents us from defining directly the function $\varphi_0(t)$ for a given stream using the usual rules for the calculation of conditional probabilities. Therefore this function will be given another, more complex definition which will determine it uniquely for any stationary stream. Further, we shall define not one function, but a whole sequence $\varphi_k(t)$ $(k = 0,1,2, \ldots)$ of functions which will be called *Palm's functions* and which will be useful later for solving a series of important problems.

Suppose that we have two consecutive periods of time, of which the first has length $\tau$ and the second $t$ (hereafter for the sake of brevity these periods will be called respectively "period $\tau$" and "period $t$"). For a given stationary stream $H_k(\tau,t)$ $(k \geqslant 0)$ will be used to denote the probability that both of the following events occur: (1) in period $\tau$ there will occur at least one call; (2) in period $t$ there will not occur more than $k$ calls. These two occurrences will, generally speaking, be interdependent. Since the probability of (1) in our former terminology is $w(\tau)$, the ratio

$$\frac{H_k(\tau,t)}{w(\tau)} \tag{9.1}$$

expresses the conditional probability of event (2) on condition that event (1) has taken place, i.e. the probability of the occurrence of not more than $k$ calls in the period $t$ on condition that in period $\tau$ there occurred at least one call.

If, as $\tau \to 0$ (and for constant $t$), this ratio tends to some limit, then it is natural to call this limit the conditional probability of the occurrence of not more than $k$ calls in period $t$ *on condition that at the first moment of this period a call occurred.*

It will now be established that as $\tau \to 0$ (and for constant $t$) the limit of the ratio (9.1) always exists, provided the given stationary stream has a finite parameter $\lambda$. To do this, consider first the ratio $H_k(\tau,t)/\tau$. To show the existence of the limit of this ratio as $\tau \to 0$, it is sufficient to show that the magnitude $H_k(\tau,t)$ as a function of $\tau$ satisfies all requirements of lemma § 7. That this function is non-negative and monotonic is self-evident. Suppose that $\tau = \tau_1 + \tau_2$ and that the period $\tau_1$ precedes the period $\tau_2$, then if the above two events occur (the joint probability of which was denoted by $H_k(\tau,t)$), clearly at least one of the following two events occurs —

(*A*) In the period $\tau_2$ there occurs at least one call, and in the period $t$ there occur not more than $k$ calls (the probability of event (*A*) is equal to $H_k(\tau_2,t)$).

(*B*) In the period $\tau_1$ there occurs at least one call, and in the period $\tau_2 + t$ there occur not more than $k$ calls (the probability of event (*B*) is equal to $H_k(\tau_1,\tau_2+t) \leqslant H_k(\tau_1,t)$, since for a fixed $\tau$, $H_k(\tau,t)$ is clearly a non-increasing function of $t$).

Thus we find

$$H_k(\tau_1+\tau_2,t) \leqslant H_k(\tau_1,t) + H_k(\tau_2,t),$$

i.e. the function $H_k(\tau,t)$ satisfies (with regard to $\tau$) even the last requirement of § 7.

Employing this lemma, it follows that as $\tau \to 0$ the ratio $H_k(\tau,t)/\tau$ tends to some limit or increases indefinitely. However, the latter possibility is excluded as obviously $H_k(\tau,t) \leqslant w(\tau)$ always, and the ratio $w(\tau)/\tau$ according to our hypothesis tends to a finite limit $\lambda$.

Finally,

$$\frac{H_k(\tau,t)}{w(\tau)} = \frac{H_k(\tau,t)/\tau}{w(\tau)/\tau}$$

and it has been shown that as $\tau \to 0$ the numerator and denominator of this fraction tend to fixed limits. Thus even

$$\lim_{\tau \to 0} \frac{H_k(\tau,t)}{w(\tau)} = \Phi_k(t) \tag{9.2}$$

exists. This limit is of course a function of $t$.

Now, suppose that

$$h_0(\tau,t) = H_0(\tau,t), \quad h_k(\tau,t) = H_k(\tau,t) - H_{k-1}(\tau,t) \qquad (k > 0).$$

Clearly, $h_k(\tau,t)$ is the probability that (1) in the period $\tau$ there is at least one call and (2) in the period $t$ there are precisely $k$ calls. The ratio $h_k(\tau,t)/w(\tau)$ correspondingly represents the conditional probability that there are $k$ calls in the period $t$, assuming that at least one call occurs in the period $\tau$. From (9.2) it follows that

$$\lim_{\tau \to 0} \frac{h_0(\tau,t)}{w(\tau)} = \Phi_0(t), \quad \lim_{\tau \to 0} \frac{h_k(\tau,t)}{w(\tau)} = \Phi_k(t) - \Phi_{k-1}(t) \qquad (k > 0).$$

Setting

$$\varphi_0(t) = \Phi_0(t), \quad \varphi_k(t) = \Phi_k(t) - \Phi_{k-1}(t) \qquad (k > 0),$$

we have

$$\lim_{\tau \to 0} \frac{h_k(\tau,t)}{w(\tau)} = \varphi_k(t) \qquad (k = 0,1,2,\ldots).$$

The functions $\varphi_k(t)$ will be called *Palm's functions*. The function $\varphi_k(t)$ can be interpreted as *the probability that k calls occur in a period of length t, assuming that a call has occurred at the first moment of this period.* Thereby it differs from the function $v_k(t)$, representing the probability of the same event when nothing is known of the corresponding first moment. The argument that has been employed shows that the whole set of Palm's functions is uniquely determined for any stationary stream with a finite parameter $\lambda$.

It should be noted that, since $H_k(\tau,t)$ is a non-increasing function of $t$ and $w(\tau)$ does not depend on $t$, so all functions $\Phi_k(t)$, and in particular the function $\Phi_0(t) = \varphi_0(t)$, are non-increasing in the region $0 < t < +\infty$.

## 10 Palm's formulae

Palm's functions are connected with the basic functions $v_k(t)$ of a given stationary stream by simple and important formulae which will now be deduced.

Suppose that a given stationary stream is orderly and has a finite parameter $\lambda$. Consider again a period of time of length $\tau + t$ comprising a "period $\tau$" and immediately following it a "period $t$". Denote the number of calls in periods $\tau$ and $t$ ($n_1$, $n_2$ are random variables) by $n_1$ and $n_2$ respectively. It follows that

$$v_k(\tau+t) = P\{n_1 + n_2 = k\}$$
$$= \sum_{r=0}^{k} P\{n_1 = r, \quad n_2 = k - r\},$$

from which, by virtue of the orderliness of the given stream, as $\tau \to 0$,

$$v_k(\tau+t) = P\{n_1 = 0, n_2 = k\} + P\{n_1 = 1, n_2 = k - 1\} + o(\tau). \quad (10.1)$$

But

$$P\{n_1 = 0, n_2 = k\} = P\{n_2 = k\} - P\{n_1 > 0, n_2 = k\} = v_k(t) - h_k(\tau,t), \quad (10.2)$$

where the terminology of § 9 is used. Further

$$P\{n_1 = 1, n_2 = k - 1\} = P\{n_1 > 0, n_2 = k - 1\} - P\{n_1 > 1, n_2 = k - 1\} = h_{k-1}(\tau,t) + o(\tau). \quad (10.3)$$

Inserting (10.2) and (10.3) into (10.1), we find

$$v_k(t+\tau) = v_k(t) - h_k(\tau,t) + h_{k-1}(\tau,t) + o(\tau)$$

from which

$$\frac{v_k(\tau+t) - v_k(t)}{\tau} = \frac{h_{k-1}(\tau,t)}{w(\tau)} \frac{w(\tau)}{\tau} - \frac{h_k(\tau,t)}{w(\tau)} \frac{w(\tau)}{\tau} + o(1).$$

By virtue of the results of § 9, this equation implies the differentiability of the function $v_k(t)$ and the following equation[*]

$$v_k'(t) = \lambda[\varphi_{k-1}(t) - \varphi_k(t)] \qquad (k > 0). \tag{10.4}$$

When $k = 0$ this same argument gives

$$v_0'(t) = -\lambda\,\varphi_0(t) \tag{10.5}$$

so that the relation (10.4) is true even for $k = 0$ if one sets $\varphi_{-1}(t) = 0$. Adding the equations (10.4) for $k = 0,1,\ldots,m$ and denoting by

$$V_m(t) = \sum_{k=0}^{m} v_k(t)$$

the probability that in the period of length $t$ there will *not be more* than $m$ calls, it follows that

$$V_m'(t) = -\lambda\,\varphi_m(t) \qquad (m = 0,1,2\ldots). \tag{10.6}$$

Formulae (10.4) and (10.6) were also the objects of our deduction (Palm has only the formula (10.5)).

It is sometimes more convenient to express these formulae in an integral form. Integrating both sides of (10.6) from 0 to $t$, we find that

$$V_m(+0) - V_m(t) = \lambda \int_0^t \varphi_m(u)du \qquad (m = 0,1,2\ldots).$$

But $\quad V_m(+0) = 1.$

This follows, since

$$V_m(+0) \geqslant V_0(+0) = v_0(+0) = 1 - w(+0),$$

and, as $t \to 0$, $w(t)/t \to \lambda$ so $w(+0) = 0$.

Thus, for any $t > 0$, it follows that

$$1 - V_m(t) = \lambda \int_0^t \varphi_m(u)du \qquad (m = 0,1,2\ldots) \tag{10.7}$$

from which it is easily seen that

$$\left.\begin{aligned} v_0(t) &= 1 - \lambda \int_0^t \varphi_0(u)du \\ v_k(t) &= \lambda \int_0^t [\varphi_{k-1}(u) - \varphi_k(u)]du \qquad (k = 1,2,\ldots). \end{aligned}\right\} \tag{10.8}$$

These formulae express simply and directly the functions $v_k(t)$ of a given stream in terms of the functions $\varphi_k(t)$ of Palm.

---

[*] See Additional Note 1 on page 121.

## 11 The intensity of a stationary stream — Korolyuk's theorem

In § 4 the intensity, $\mu$, of a given stationary stream was defined as the mathematical expectation of the number of calls in a unit of time. By virtue of the additivity of mathematical expectations, it was then found that the mathematical expectation of the number of calls in a period of length $t$ for a stationary stream was proportional to $t$, i.e.

$$\sum_{k=0}^{\infty} kv_k(t) = \mu t.$$

It was then established that $\mu \geqslant \lambda$ always, and for a simple stream $\mu = \lambda$.

In works of an applied nature the coincidence of parameters $\mu$ and $\lambda$ is usually accepted as a self-evident fact, not even requiring discussion in the investigation of stationary streams of the most general type. In view of the practical significance of this assumption, it seems important to clear up its presuppositions and give it a formal basis whenever possible.

First let us look at the case of a stationary stream without after-effects. In § 8 it was shown that for a stream of this kind the generating function

$$F(t,x) = \sum_{k=0}^{\infty} v_k(t)x^k$$

has the form

$$\exp \lambda t[\Phi(x) - 1]$$

where

$$\lambda > 0,\ \Phi(x) = \sum_{l=1}^{\infty} p_l\, x^l, \quad p_l \geqslant 0\ (l = 1,2\ldots), \quad \sum_{l=1}^{\infty} p_l = 1.$$

Since

$$\mu t = \sum_{k=0}^{\infty} kv_k(t) = \left\{\frac{\partial F(t,x)}{\partial x}\right\}_{x=1}.$$

and since

$$\frac{\partial F(t,x)}{\partial x} = F(t,x)\lambda t\Phi'(x), \quad F(t,1) = \sum_{k=0}^{\infty} v_k(t) = 1$$

it follows that

$$\mu t = \lambda t\Phi'(1) = \lambda t[p_1 + 2p_2 + 3p_3 + \ldots],$$

whence

$$\mu = \lambda[p_1 + 2p_2 + 3p_3 + \ldots].$$

Further, since

$$\sum_{l=1}^{\infty} p_l = 1$$

and

$$p_l \geqslant 0 \qquad (l \geqslant 1)$$

it follows that for the equation $\mu = \lambda$ to hold, it is necessary and sufficient to have $p_1 = 1$. But when $p_1 = 1$ the given stream, as was seen at the end of § 8, is simple. Thus *among stationary streams without after-effects only simple streams satisfy the requirement $\mu = \lambda$; for the remainder $\mu > \lambda$.*

Since orderliness is a necessary and sufficient condition for a stationary stream without after-effects to be simple, it is possible to say further that *for a stationary stream without after-effects its orderliness is a necessary and sufficient condition for the equation $\mu = \lambda$ to hold.*

Palm's formulae which were derived in § 10 allow us to establish quite easily that, as V. S. Korolyuk has shown, for any stationary stream orderliness implies the equation $\mu = \lambda$ (where the case $\mu = \lambda = +\infty$ is not excluded).

In fact, we have

$$\mu = \sum_{k=1}^{\infty} k v_k(1) = \sum_{k=1}^{\infty} [v_k(1) + v_{k+1}(1) + v_{k+2}(1) + \ldots]$$
$$= \sum_{k=1}^{\infty} [1 - V_{k-1}(1)] = \sum_{k=0}^{\infty} [1 - V_k(1)]$$

from which, by virtue of formula (10.7),

$$\mu = \lambda \sum_{k=0}^{\infty} \int_0^1 \varphi_k(u) du. \tag{11.1}$$

But

$$\varphi_k(u) = \lim_{\tau \to 0} \frac{h_k(\tau,u)}{w(\tau)} \tag{11.2}$$

and as the ratio $h_k(\tau,u)/w(\tau)$ is the conditional probability of there occurring precisely $k$ calls in the period $u$ (on condition that there are calls in period $\tau$), so for any $m > 0$

$$\sum_{k=0}^{m} \frac{h_k(\tau,u)}{w(\tau)} \leqslant 1 \qquad (0 < u \leqslant 1)$$

and, therefore, by virtue of (11.2)

$$\sum_{k=0}^{m} \varphi_k(u) \leqslant 1 \qquad (0 < u \leqslant 1).$$

Consequently for any $m > 0$

$$\sum_{k=0}^{m} \int_0^1 \varphi_k(u) du = \int_0^1 \left\{ \sum_{k=0}^{m} \varphi_k(u) \right\} du \leqslant 1$$

and this implies that

$$\sum_{k=0}^{\infty} \int_0^1 \varphi_k(u) du \leqslant 1$$

and (11.1) gives $\mu \leqslant \lambda$. Further, as was seen in § 4, $\mu \geqslant \lambda$ always, and hence, here, $\mu = \lambda$ and our assertion is proved.

CHAPTER 4

# STREAMS WITH LIMITED AFTER-EFFECTS

## 12 Another method of describing a stream of calls

The method of presenting a stream of calls described in § 6 arises from understanding it as a random process $x(t)$ and does not differ from the general method of describing any random process. This automatic inclusion of the theory of streams of calls in the general theory of random processes doubtless has its advantages as it allows us to bring to its study methods and results of the general theory of random processes. However, taking into account the specific properties of our incoming streams as random processes (and first of all that the magnitude $x(t)$ is always monotonic and bears only integer non-negative values), we can hope to find a means of describing them which, though less general, may be more simple and convenient.

Let us suppose again that the first call of a given incoming stream occurs at $t_0 = 0$; suppose $t_i$ $(i = 1,2, \dots)$ is the time of call $i$, so that $t_{i-1} \leqslant t_i$ $(i = 1,2, \dots)$. Finally let

$$t_i - t_{i-1} = z_i \qquad (i = 1,2, \dots)$$

so that $z_1 = t_1$ and, when $i > 1$, $z$ denotes the length of the period between the $(i - 1)$th and $i$th calls. Clearly, all $t_i$ and $z_i$ $(i = 1,2, \dots)$ represent random variables capable of bearing only non-negative values.

A stream may be regarded as given, if for any $n > 0$ the $n$-dimensional distribution function of the vector $(z_1, z_2, \dots, z_n)$ is given. Clearly this method of describing a stream is more elementary than the one employed in § 6 since at the basis of it lies the presentation of the stream not as a random process of general type but as a sequence of random variables. It will now be established that the two methods of describing a stream are equivalent, i.e. that a stream given by means of either of these two methods will be determined equally well by the other.

Suppose, as in § 6, $x(t)$ denotes the number of calls before the moment $t$. Clearly the inequalities $t_k < u$ and $x(u) \geqslant k$ express one and the same thing. The same equivalence exists for the inequalities $t_k \geqslant u$, $x(u) < k$ and for the inequalities $u \leqslant t_k < v$, $x(u) < k \leqslant x(v)$. Thus it follows that the system of inequalities

$$u_k \leqslant t_k < v_k \qquad (1 \leqslant k \leqslant n) \tag{12.1}$$

expresses the same things as the system of inequalities

$$x(u_k) < k \leqslant x(v_k) \qquad (1 \leqslant k \leqslant n) \tag{12.2}$$

whatever the real numbers $u_k$, $v_k$ $(1 \leqslant k \leqslant n)$.

If the stream of calls is given in the sense of § 6, then for any $n$ and for any $u_k$, $v_k$ $(1 \leqslant k \leqslant n)$ the distribution function of the $2n$-dimensional vector $[x(u_k), x(v_k); k = 1,2, \ldots n]$ is given and the probability of system (12.2) is uniquely defined, and consequently the system (12.1) which is equivalent to it. But in view of the arbitrariness of the numbers $u_k$, $v_k$ this latter statement implies that the distribution function of the vector $(t_1, t_2, \ldots, t_n)$ has been uniquely given, and hence also of the vector

$$(z_1, z_2, \ldots, z_n) = t_1, t_2 - t_1, \ldots, t_n - t_{n-1}).$$

Thus the given stream is also uniquely defined in a new sense.

Conversely if the distribution function of vector $(z_1, z_2, \ldots, z_n)$ is given for any $n$, then since

$$t_k = \sum_{i=1}^{k} z_i \qquad (1 \leqslant k \leqslant n)$$

the distribution function of the vector $(t_1, t_2, \ldots, t_n)$ is given also. But for any $u_1, u_2 \ldots u_n$ and any integers $k_1, k_2, \ldots, k_n$, the system of inequalities

$$t_{k_i} < u_i, \; t_{k_{i+1}} \geqslant u_i \qquad (1 \leqslant i \leqslant n) \tag{12.3}$$

is equivalent to the system of equalities

$$x(u_i) = k_i \qquad (1 \leqslant i \leqslant n), \tag{12.4}$$

and as the probability of system (12.3) has been uniquely defined, the same is true for system (12.4) also. This means that for any $n$ and any $u_i$ $(1 \leqslant i \leqslant n)$ the distribution function of the vector $[x(u_1), x(u_2), \ldots, x(u_n)]$ is uniquely defined, i.e. that our process has been uniquely defined in the sense of § 6.

Thus this new method of describing a stream is actually equivalent to the one employed in § 6.

## 13 Streams with limited after-effects

If a given stream is without after-effects then the variables $z_1, z_2, \ldots, z_n$ are mutually independent. However, the reverse conclusion, as will be shown later, would not be true. The mutual independence of variables $z_k$ limits to a significant degree the occurrence of after-effects, but does not exclude it completely. In Part II it will be shown that it is precisely the streams with mutually independent $z_k$ but with after-effects that play an important role in the theory of service of calls by a fully accessible collection of lines. It is therefore necessary now to establish certain basic properties of such streams.

Following Palm, any stream in which $z_1, z_2, \ldots, z_n$ is a sequence of mutually independent random variables will be called a *stream with limited after-effects*. Clearly, for the unique description of any stream, it is sufficient to give the distribution functions of all the variables $z_k$ $(k = 1,2, \ldots)$. Hereafter these laws will be denoted by $F_k(x)$ $(k = 1,2, \ldots)$.

If the given stream is stationary and orderly, a complete absence of after-effects implies that the stream is simple, which type was studied in detail in Chapter 1. Thus a stationary and orderly stream with a limited after-effect may be regarded as a certain generalisation of the simple stream(*). Streams of this type are of special interest in the theory of service for the case of systems with losses (*vide* Chapter 8). For brevity, a stationary orderly stream with a limited after-effect will be called a *stream of type P* (or Palm stream).

In § 9, the system of "Palm's functions" $\varphi_k(t)$ $(k = 0,1,2\ldots)$ was defined for any stationary stream. The function $\varphi_0(t)$, as will now be seen, plays a fundamental role in the theory of streams of type $P$. *A stream of type P is uniquely defined by this function.* Actually, since for a stream of type $P$ the variables $z_1, z_2, \ldots, z_n$ are mutually independent, it follows that for the unique determination of the distribution function of each vector $(z_1, z_2, \ldots, z_n)$ $(n = 1,2,\ldots)$, and for the unique description of the stream, it is sufficient to give the distribution functions $F_k(x)$ of the variables $z_k$ $(k = 1,2,\ldots)$. But these functions are uniquely determined by specifying Palm's function $\varphi_0(t)$, as the following proposition shows.

*Theorem. For a stream of type P*

$$F_1(x) = \lambda \int_0^x \varphi_0(u)du, \quad F_k(x) = 1 - \varphi_0(x) \qquad (k \geqslant 2).$$

*Proof.* For the best presentation the proof will be divided into several stages.

(1) Since $F_1(x) = P\{t_1 < x\}$ is the probability of the occurrence of calls in the period $(0,x)$, and consequently, in our old terminology, is equal to

$$w(x) = 1 - v_0(x) = 1 - V_0(x),$$

the stated expression for $F_1(x)$ follows directly from the formula (10.7) when $m = 0$. When this is so the parameter $\lambda$ of a given stream is determined from the function $\varphi_0(t)$, using the relation

$$\lambda \int_0^\infty \varphi_0(u)du = F_1(+\infty) = 1.$$

In like fashion, the instance $k > 1$ is next considered.

(2) Suppose, as in § 8, that

$$\psi_k(t) = 1 - V_{k-1}(t) = v_k(t) + v_{k+1}(t) + \ldots.$$

(*) It is interesting to note that the limitedness of the after-effect follows from the absence of the after-effect only for orderly streams. A non-orderly stream without after-effect is not necessarily a stream with a limited after-effect. The stream investigated at the end of § 8 can be regarded as such an example when $p_1 = p_2 = \frac{1}{2}$ (for this observation the author is indebted to P. I. Vassilyev).

Then there follows the

*Lemma. For any stream of type P, and for any $r > 0$,*

$$\frac{\psi_{r+1}(u)}{\psi_r(u)} \to 0 \qquad (u \to 0).$$

*Proof.* Denoting by $t_k$ the moment of the $k$th call and assuming, as before, that $t_k - t_{k-1} = z_k$ $(k = 1,2, \ldots)$, we have

$$\psi_{r+1}(u) = P\{t_{r+1} < u\} \leqslant P\{t_r < u,\ z_{r+1} < u\},$$

and consequently, since $z_{r+1}$ is independent of $t_r$,

$$\psi_{r+1}(u) \leqslant \psi_r(u) F_{r+1}(u).$$

To prove the lemma, it is therefore sufficient to establish that $F_{r+1}(u) \to 0$ as $u \to 0$.

Suppose that $a > 0$ is so great that $\psi_r(a) > 0$.

Let $x > 0$ be arbitrarily small and $n$ an integer such that $(n - 1)x < a \leqslant nx$. The segments $[(k-1)x,\ kx]$ $(1 \leqslant k \leqslant n)$ will be called "compartments". If $t_r < nx$ and $z_{r+1} < x$, then the moments $t_r$ and $t_{r+1}$ lie either in one compartment or in two neighbouring compartments, so that at least one out of the $n$ segments

$$[(l-1)x,\ (l+1)x] \qquad (l = 1,2, \ldots, n)$$

of length $2x$ contains more than one call. Then

$$P\{t_r < nx,\ z_{r+1} < x\} = \psi_r(nx) F_{r+1}(x) \leqslant n\psi_2(2x)$$
$$= 2nx\,\frac{\psi_2(2x)}{2x} < 2(a+x)\,\frac{\psi_2(2x)}{2x}$$

and when $x < a$

$$F_{r+1}(x) \leqslant \frac{2(a+x)}{\psi_r(nx)}\,\frac{\psi_2(2x)}{2x} \leqslant \frac{4a}{\psi_r(a)}\,\frac{\psi_2(2x)}{2x} \to 0$$

as $x \to 0$ by virtue of the orderliness of the given stream. This completes the proof of our lemma.

(3) Turning now to the proof of the theorem, we first establish that $F_2(t) = 1 - \varphi_0(t)$. For this purpose, consider again the probability introduced in Chapter 3, the probability $h_0(\tau,t)$ that in a certain period of length $\tau$, calls occur, and that in the period of length $t$ following it there are none. If the number of calls in the period $\tau$ is equal to $k$, then it follows from the absence of calls in period $t$ that $z_{k+1} > t$ and the probability of this is $1 - F_{k+1}(t)$.

Thus

$$h_0(\tau,t) \leqslant \sum_{k=1}^{\infty} v_k(\tau)\,[1 - F_{k+1}(t)] \leqslant v_1(\tau)\,[1 - F_2(t)] + \psi_2(\tau).$$

On the other hand, for the same event ($k$ calls in period $\tau$) it follows from $z_{k+1} > t + \tau$ that in the period $t$ there are no calls. Therefore,

$$h_0(\tau,t) \geqslant \sum_{k=1}^{\infty} v_k(\tau)\,[1 - F_{k+1}(t+\tau)] \geqslant v_1(\tau)\,[1 - F_2(t+\tau)]$$

and

$$v_1(\tau)\,[1 - F_2(t+\tau)] \leqslant h_0(\tau,t) \leqslant v_1(\tau)\,[1 - F_2(t)] + \psi_2(\tau).$$

Divide each part of this inequality by $w(\tau)$ and noting that, as $\tau \to 0$,

$$\frac{v_1(\tau)}{w(\tau)} \to 1, \quad \frac{\psi_2(\tau)}{w(\tau)} \to 0, \quad \frac{h_0(\tau,t)}{w(\tau)} \to \varphi_0(t),$$

we find in the limit,

$$1 - F_2(t+0) \leqslant \varphi_0(t) \leqslant 1 - F_2(t);$$

whence $F_2(t) = 1 - \varphi_0(t)$ at all points of continuity of the distribution function $F_2(t)$.

(4) Now it will be established by induction that

$$F_{r+2}(t) = 1 - \varphi_0(t) \text{ for all } r \geqslant 0.$$

By virtue of the stationarity of the given stream, the distribution function of the distance between the first two calls following any moment $a > 0$ equals the distribution function $F_2(t)$ of the distance $z_2$ between the first two calls following the moment 0. But if there are $r$ calls in the period $(0,a)$ the distance between the first two calls following the moment $a$ is $z_{r+2}$, and its distribution function, $F_{r+2}(t)$, does not depend on the preceding course of the stream. Thus

$$F_2(t) = \sum_{r=0}^{\infty} v_r(a) F_{r+2}(t). \tag{13.1}$$

Now, suppose that it is already established that

$$F_2(t) = F_3(t) = \ldots = F_{r+1}(t) = 1 - \varphi_0(t).$$

Then (13.1) gives

$$1 - \varphi_0(t) = [1 - \varphi_0(t)] \sum_{k=0}^{r-1} v_k(a) + v_r(a) F_{r+2}(t) + \sum_{k>r} v_k(a) F_{k+2}(t),$$

from which, by virtue of $v_r(a) = \psi_r(a) - \psi_{r+1}(a)$

$$[1 - \varphi_0(t)]\,\psi_r(a) - F_{r+2}(t)\,\psi_r(a) = -\,\psi_{r+1}(a) F_{r+2}(t) + \sum_{k>r} v_k(a) F_{k+2}(t).$$

As the last sum of the right-hand side does not exceed

$$\sum_{k > r} v_k(a) = \psi_{r+1}(a),$$

it follows that(*)

$$\psi_r(a)\,|1 - \varphi_0(t) - F_{r+2}(t)| \leqslant \psi_{r+1}(a);$$

$$|1 - \varphi_0(t) - F_{r+2}(t)| \leqslant \frac{\psi_{r+1}(a)}{\psi_r(a)}.$$

This inequality is true for any $t > 0$ and $a > 0$. But as $a \to 0$ the right-hand side tends to zero, by the lemma which we have proved. Since the left-hand side of the inequality does not depend on $a$, for any $t > 0$

$$F_{r+2}(t) = 1 - \varphi_0(t)$$

which was to be proved.

(*)See Additional Note 2 on page 121.

CHAPTER 5

# LIMIT THEOREM

## 14 Statement of the problem — Palm's theorem

It was stated in Chapter 1 that the great majority of investigations of an applied character are based on the assumption that the primitive stream of calls entering a given establishment is of simple type. However it has been known for some time that there is a whole series of theoretical considerations which, in most practical instances, cast doubt upon the accuracy of the assumptions employed in describing the simple stream. (This is particularly so with regard to the requirement of the absence of after-effects.) Thus one must not be surprised if observation and experience of processes actually encountered reveal a certain minor deviation from processes of simple type. Surprise may be more occasioned by the fact that deviations of this kind are in the majority of cases less significant than might have been expected from theoretical considerations. Thus, although an investigator, when applying a theory to experimental data, usually has the task of explaining the causes of deviation of actual processes from the course theoretically predicted for them, the situation is in fact the wrong way round — experimental data agree with the deductions from the constructed theory, as a rule, better than could have been expected on the principal considerations, and it is precisely this "much better" agreement which requires explanation.

Palm [8] has made a noteworthy attempt to explain facts of this kind by the assumption that a given process represents a simple sum (superposition) of a large number of mutually independent processes of small intensity, in which each of these processes is stationary and orderly and can be either with or without after-effects. In this it is shown that under very broad assumptions the summary stream must be very near to a simple one. Such a statement of the problem is, on the face of it, very near to the real situation. So, if a large number of subscribers are attached to the given exchange, a general stream of calls is composed of streams (of a comparatively small intensity) arising from different subscribers, in which the component streams can in the first instance be regarded as stationary, orderly, and mutually independent.

In this way we arrive at a series of original limit theorems capable of explaining to a significant degree the resultant phenomenon. We shall deal with this question in the present chapter.

Suppose the stream under consideration represents the superposition of $n$ stationary, orderly, and mutually independent streams. $\lambda_r$ will denote the intensity of the $r$th stream, $\phi_r(t)$ its Palm function (which we denoted by $\phi_0(t)$ in Chapter 3), and $v_{kr}(t)$ the probability of the occurrence in period $(0,t)$ of $k$ calls of the $r$th stream. The same variables for the total stream will be denoted respectively by $\Lambda$, $\varphi(t)$ and $V_k(t)$ (so that in particular $\Lambda = \lambda_1 + \lambda_2 \ldots + \lambda_n$). We will start from the following assumptions —

(1) As $n \to \infty$, $\Lambda$ remains a constant while the numbers $\lambda_1, \lambda_2, \ldots \lambda_n$ all tend to zero, so that for any $\varepsilon > 0$ we have $\lambda_r < \varepsilon$ ($r = 1,2, \ldots, n$), if $n$ is sufficiently large.

(2) For any constant $t > 0$ and as $n \to \infty$ the numbers $\varphi_r(t)$ ($r = 1,2, \ldots, n$) all tend to unity, so that for any $\varepsilon > 0$ we have $1 - \varphi_r(t) < \varepsilon (1 \leqslant r \leqslant n)$ if $n$ is sufficiently large.

The second assumption requires some explanation. The closest analysis shows that the uniform decreases of the intensities of the component streams expressed by assumption (1) are not a sufficient condition for the total stream to approach a simple one.

An accumulation of a large number of calls in small sections of one and the same stream can be responsible for this — an accumulation the possibility of which is created by the fact that the after-effect in each of the component streams has not hitherto been subject to any limit. Assumption 2 has the object of diminishing the chances of this kind of accumulation. It says that, for $t$ however large, the probability of not receiving one new call *from the same stream* after a certain call during time $t$ should, as $n \to \infty$, tend to unity, equally in all component streams.

First, we shall show that under the above assumptions the probability $V_0(t)$ of an absence of calls in the total stream in period $(0,t)$ approaches as $n \to \infty$ to the corresponding probability for a simple stream with parameter $\Lambda$.

*Palm's Theorem. For a constant* $t > 0$ *and as* $n \to \infty$,

$$V_0(t) \to e^{-\Lambda t}.$$

*Proof.* The formula gives for the $r$th stream

$$1 - v_{0r}(t) = \lambda_r \int_0^t \varphi_r(u)du$$

$$= \lambda_r t - \lambda_r \int_0^t [1 - \varphi_r(u)]du,$$

or

$$v_{0r}(t) = 1 - \lambda_r t + \lambda_r \int_0^t [1 - \varphi_r(u)]du.$$

Thus, by virtue of assumption (2), for a sufficiently large $n$,

$$v_{0r}(t) = 1 - \lambda_r t + \varepsilon \theta_r \lambda_r t, \quad |\theta_r| < 1, \quad (1 \leqslant r \leqslant n) \quad (14.1)$$

from which we find

$$|\log_e v_{0r}(t) + \lambda_r t| < c(t)\varepsilon\lambda_r \qquad (1 \leqslant r \leqslant n),$$

where $c(t) > 0$ depends only on $t$. Since, by virtue of the mutual independence of the streams

$$V_0(t) = \prod_{r=1}^{n} v_{0r}(t),$$

it follows that

$$|\log_e V_0(t) + \Lambda t| = \left| \sum_{r=1}^{n} [\log_e v_{0r}(t) + \lambda_r t] \right|$$

$$\leqslant \left| \sum_{r=1}^{n} |\log_e v_{0r}(t) + \lambda_r t| \right| < c(t)\varepsilon\Lambda.$$

Since $\varepsilon > 0$ is arbitrarily small for a sufficiently large $n$, so, as $n \to \infty$,

$$\log_e V_0(t) \to -\Lambda t, \quad V_0(t) \to e^{-\Lambda t},$$

which was to be proved.

Without sufficient grounds, Palm supposes that the theorem proved already implies a total stream having approximately a simple character(*). Of course this theorem is nowhere near sufficient, and it is now necessary to investigate this question further.

## 15 Limiting behaviour of functions

We shall first establish that for any $k > 0$ the function $V_k(t)$ of our total stream tends as $n \to \infty$ to the corresponding function of a simple stream with parameter $\Lambda$, i.e. to $e^{-\Lambda t}(\Lambda t)^k/k!$. For this purpose we require the following lemma.

*Lemma. Suppose we have a stationary and orderly stream with intensity $\lambda$ and the Palm function $\varphi(t)$ and suppose that $\psi(t)$ denotes as before the probability of the occurrence of not less than two calls during time $t$. Then, for any $t > 0$,*

$$\psi(t) \leqslant \lambda t\,[1 - \varphi(t)].$$

*Proof.* Let us divide the segment $(0,t)$ into $m$ equal parts (compartments)

$$\Delta_k = \left(\frac{k-1}{m}\,t, \frac{k}{m}\,t\right) \qquad (1 \leqslant k \leqslant m).$$

(*) We should observe at this point that Palm's proof imposes superfluous requirements on the function $\phi_r(t)$ on the one hand, and on the other has various lacunae.

The occurrence in segment $(0,t)$ of at least two calls clearly implies the occurrence of at least one of the following two instances —

($A$) There exists at least one compartment $\Delta_k$ containing not less than two calls.

($B$) There exists a compartment $\Delta_k$ ($< m$) such that calls occur both in $\Delta_k$ and in the segment $\left(\frac{k}{m}\, t,t\right)$.

Thus we have

$$\psi(t) \leqslant P(A) + P(B). \tag{15.1}$$

First of all we have by virtue of the orderliness of the given stream

$$P(A) \leqslant m\psi\left(\frac{t}{m}\right) = t\,\frac{\psi\left(\frac{t}{m}\right)}{\frac{t}{m}} \to 0 \qquad (m \to \infty) \tag{15.2}$$

Further in § 9 we denoted by $h_0(\tau,t)$ the probability that in a period of length $\tau$ there were calls but in the period of length $t$ following it there were none. Therefore, $w(\tau) - h_0(\tau,t)$ is the probability that there are calls in $\tau$ and in $t$. Thus

$$P(B) \leqslant \sum_{k=1}^{m-1}\left[w\left(\frac{t}{m}\right) - h_0\left(\frac{t}{m}, \frac{m-k}{m}\,t\right)\right]$$

$$\leqslant m\left[w\left(\frac{t}{m}\right) - h_0\left(\frac{t}{m}, t\right)\right] = mw\left(\frac{t}{m}\right)\left[1 - \frac{h_0\left(\frac{t}{m}, t\right)}{w\left(\frac{t}{m}\right)}\right] \tag{15.3}$$

Since, as $m \to \infty$,

$$mw\left(\frac{t}{m}\right) = t\,\frac{w\left(\frac{t}{m}\right)}{\frac{t}{m}} \to \lambda t,$$

so the right-hand side of the last inequalities tends to $\lambda t[1 - \varphi(t)]$ as $m \to \infty$. Further, since $\psi(t)$ does not depend on $m$, so from (15.1), (15.2) and (15.3), it follows that in the limit, as $m \to \infty$,

$$\psi(t) \leqslant \lambda t[1 - \varphi(t)],$$

which was to be proved.

The following terminology will hereafter be used —

$A_k$ means that in the period $(0,t)$ there occur $k$ calls of the total stream;

$H_1$ means that none of the given streams gives more than one call in $(0,t)$;

$H_2$ means that at least one of the given streams gives more than one call in $(0,t)$.

Our objective is the investigation of the asymptotic behaviour of the magnitude $V_k(t) = P(A_k)$. But

$$P(A_k) = P(H_1A_k) + P(H_2A_k)$$

and, denoting the function $\psi(t)$ for the $r$th component stream by $\psi_r(t)$, we find by virtue of the lemma which we have proved that

$$P(H_2A_k) \leqslant P(H_2) \leqslant \sum_{r=1}^{n} \psi_r(t) \leqslant \sum_{r=1}^{n} \lambda_r t[1 - \varphi_r(t)].$$

Since, by assumption (2), we have for sufficiently large $n$

$$1 - \varphi_r(t) < \varepsilon \qquad (r = 1,2, \ldots, n)$$

it follows that for sufficiently large $n$

$$P(H_2A_k) \leqslant \varepsilon t \Lambda,$$

and consequently that, as $n \to \infty$,

$$P(H_2A_k) \to 0, \; V_k(t) = P(H_1A_k) + o(1). \qquad (15.4)$$

But the event $H_1A_k$ obviously requires that out of the $n$ component streams a certain $k$ give in the period $(0,t)$ exactly one call, while the remaining $n - k$ give no calls in this period. Thus if $C$ $(r_1, r_2, \ldots, r_k)$ denotes the arbitrary combination of $k$ numbers differing among themselves in the series $1,2, \ldots, n$, then

$$P(H_1A_k) = \sum_C \frac{v_{1r_1}(t)v_{1r_2}(t) \ldots v_{1r_k}(t)}{v_{0r_1}(t)v_{0r_2}(t) \ldots v_{0r_k}(t)} \prod_{l=1}^{n} v_{0l}(t)$$

$$= V_0(t) \sum_C \prod_{p=1}^{k} \frac{v_{1r_p}(t)}{v_{0r_p}(t)}, \qquad (15.5)$$

where the summation occurs for all combinations of the specified type.

Now we can go on to the proof of our basic assertion.

*Theorem. For* $n \to \infty$,

$$V_k(t) \to e^{-\Lambda t} \frac{(\Lambda t)^k}{k!} \qquad (k = 0,1,2, \ldots).$$

*Proof.* From (14.1) it follows that for sufficiently large $n$

$$|w_r(t) - \lambda_r t| = |1 - v_{0r}(t) - \lambda_r t| < \varepsilon \lambda_r t \qquad (1 \leqslant r \leqslant n).$$

Since $v_{1r}(t) = w_r(t) - \psi_r(t)$ and since, by virtue of the lemma proved, $\psi_r(t) < \varepsilon \lambda_r t$ for sufficiently large $n$, we can write for any fixed $t$

$$\left.\begin{aligned} v_{1r_p}(t) &= \lambda_{r_p} t + q_1 \varepsilon \lambda_{r_p} t, \\ v_{0r_p}(t) &= 1 - \lambda_{r_p} t + q_2 \varepsilon \lambda_{r_p} t = 1 + q_3 \varepsilon \end{aligned}\right\} \qquad (p = 1,2, \ldots, k)$$

where $q_1$, $q_2$, $q_3$ (as also $q_4$, $q_5$ onwards) are fixed as $n \to \infty$

Thus

$$\frac{v_{1r_p}(t)}{v_{0r_p}(t)} = \frac{\lambda_{r_p}t(1+q_1\varepsilon)}{1+q_3\varepsilon} = \lambda_{r_p}t(1+q_4\varepsilon) \quad (p = 1,2, \ldots, k)$$

and consequently

$$\prod_{p=1}^{k} \frac{v_{1r_p}(t)}{v_{0r_p}(t)} = \lambda_{r_1} \lambda_{r_2} \ldots \lambda_{r_k} t^k \, (1+q_5\varepsilon).$$

By virtue of (15.5) and Palm's theorem (§ 14) it follows that

$$P(H_1 A_k) = e^{-\Lambda t} \, t^k (1+q_6\varepsilon) \sum_C \lambda_{r_1} \lambda_{r_2} \ldots \lambda_{r_k}.$$

Since $\varepsilon$ is arbitrarily small for sufficiently large $n$, by virtue of (15.4) it is sufficient for the proof of our theorem to establish that as $n \to \infty$(*)

$$S_k = \sum_{C_k} \lambda_{r_1} \lambda_{r_2} \ldots \lambda_{r_k} \to \frac{\Lambda^k}{k!}. \tag{15.6}$$

We shall now do this. When $k = 1$ the relation (15.6) is trivial. Therefore suppose that for $k > 1$, as $n \to \infty$,

$$S_{k-1} = \sum_{C_{k-1}} \lambda_{r_1} \lambda_{r_2} \ldots \lambda_{r_{k-1}} = \frac{\Lambda^{k-1}}{(k-1)!} + o(1). \tag{15.7}$$

Let us multiply each term of the sum $S_{k-1}$ by the sum of all $\lambda_i$ not entering into it, i.e. by the magnitude

$$\Lambda - \lambda_{r_1} - \lambda_{r_2} - \ldots - \lambda_{r_{k-1}}.$$

Then, after the removal of all brackets, we obtain a sum of products of the terms

$$\lambda_{r_1} \lambda_{r_2} \ldots \lambda_{r_k}$$

where all pairs of the indices $r_1, r_2, \ldots, r_k$ differ from each other.

Each of these new products is one of the terms of the sum $S_k$. On the other hand, any term of the sum $S_k$ can be obtained by this operation and, in fact, occurs $k$ times [the term $\lambda_{r_1} \lambda_{r_2} \ldots \lambda_{r_{k-1}} \lambda_{r_k}$ is obtained as $(\lambda_{r_1} \lambda_{r_2} \ldots \lambda_{r_{k-1}})\lambda_{r_k}$, as $(\lambda_{r_1} \lambda_{r_2} \ldots \lambda_{r_{k-2}} \lambda_{r_k})\lambda_{r_{k-1}}$ etc. and finally as $(\lambda_{r_2} \ldots \lambda_{r_k})\lambda_{r_1}$]. Since, for sufficiently large $n$, and for any combination $C_{k-1}$

$$\Lambda - (k-1)\,\varepsilon \leqslant \Lambda - \lambda_{r_1} - \lambda_{r_2} - \ldots - \lambda_{r_{k-1}} \leqslant \Lambda,$$

it follows from our calculation that

$$(\Lambda - k\varepsilon)\, S_{k-1} \leqslant kS_k \leqslant \Lambda\, S_{k-1},$$

and consequently that from (15.7), as $n \to \infty$,

$$(\Lambda - k\varepsilon)\left[\frac{\Lambda^{k-1}}{(k-1)!} + o(1)\right] \leqslant kS_k \leqslant \Lambda\left[\frac{\Lambda^{k-1}}{(k-1)!} + o(1)\right].$$

---

(*)Here, for greater clarity we shall denote the various combinations of $k$ of the numbers $1,2, \ldots, n$ by $C_k$ (instead of the earlier term $C$).

But, as $\varepsilon$ is arbitrarily small for $n$ sufficiently large, it follows that $kS_k \to \Lambda^k/(k-1)!$ as $n \to \infty$, i.e. that

$$\lim_{n\to\infty} S_k = \frac{\Lambda^k}{k!},$$

and our theorem is proved by induction.

## 16 Limit theorem

The theorem which we have just proved establishes that under the specified conditions, and as $n \to \infty$, the functions $V_k(t)$ for a total stream tend to the corresponding functions of a simple stream with parameter $\Lambda$. This, however, does not mean that our total stream itself approximates to this simple stream. The fact is that, as shown in § 6, the set of functions $V_k(t)$ $(k = 0,1,2, \ldots)$ uniquely defines a given stream only if it is a stream without after-effects. We did not investigate the question of after-effects in our total stream. Thus the question of the approximation of this total stream to a simple stream with parameter $\Lambda$ requires further study.

As the concluding formula of § 6 shows, for a simple stream with parameter $\Lambda$, a defining formula is

$$P\{x(t_i) = k_i, \quad 1 \leqslant i \leqslant m\}$$
$$= e^{-\Lambda t_m}\Lambda^{k_m}\prod_{i=1}^{m}\frac{(t_i-t_{i-1})^{k_i-k_{i-1}}}{(k_i-k_{i-1})!} \tag{16.1}$$

where $t_0 = k_0 = 0$, $m$ is any positive integer, $0 < t_1 < t_2 < \ldots < t_m$, $0 \leqslant k_1 \leqslant k_2 \leqslant \ldots \leqslant k_m$, and all $k_i$ are non-negative integers. Thus the total stream may be regarded as tending to a simple stream with parameter $\Lambda$ if for this total stream the probability on the left-hand side of the equation (16.1) for any $m$, $t_i$, $k_i$ $(1 \leqslant i \leqslant m)$ and when $n$ is increasing without bound, has as its limit the right-hand side of this equation. This will now be established.

First, let us introduce some terms more convenient for this purpose. Suppose that

$$t_i - t_{i-1} = u_i, \qquad k_i - k_{i-1} = l_i,$$
$$t_m = \sum_{i=1}^{m} u_i = u, \qquad k_m = \sum_{i=1}^{m} l_i = k$$

and denote by $n(u_i)$ the number of calls entering in the period $u_i = (t_{i-1}, t_i)$. Then clearly equation (16.1) is equivalent to the equation

$$P\{n(u_i) = l_i, 1 \leqslant i \leqslant m\} = e^{-\Lambda u}\Lambda^k \prod_{i=1}^{m}\frac{u_i^{l_i}}{l_i!}. \tag{16.2}$$

In § 15 we investigated the instances $H_1A_k$ that during a certain period of time $(0,u)$ there occur $k$ calls and that all these calls belong to different component streams. Divide the segment $(0,u)$ into $m$ parts $u_1, u_2, \ldots, u_m$ whose lengths will be denoted by the same letters, and suppose that instance $H_1A_k$ occurs. By virtue of the stationarity and mutual independence of the component streams, for any of $k$ calls occurring in the period $(0,u)$ the probability of falling within the segment $u_i$ $(1 \leqslant i \leqslant m)$ is then equal to $u_i/u$ whatever the position of this segment and at whatever moments the remaining incoming calls occur. We shall denote by $B$ the event

$$n(u_i) = l_i \qquad (1 \leqslant i \leqslant m).$$

Then from the above it follows that

$$P_{H_1A_k}(B) = \frac{k!}{l_1!\, l_2! \dots l_m!} \left(\frac{u_1}{u}\right)^{l_1} \left(\frac{u_2}{u}\right)^{l_2} \dots \left(\frac{u_m}{u}\right)^{l_m}.$$

But, in § 15 we proved that, as $n \to \infty$,

$$V_k(u) = P(A_k) \to e^{-\Lambda u} \frac{(\Lambda u)^k}{k!},$$

$$P(H_1A_k) \to e^{-\Lambda u} \frac{(\Lambda u)^k}{k!}, \qquad P(H_2A_k) \to 0.$$

Thus, acknowledging that, since $\sum_{i=1}^{m} l_i = k$, the instance $A_k$ is a particular case of instance $B$, we find, as $n \to \infty$,

$$\begin{aligned} P(B) &= P(A_kB) = P(H_1A_kB) + P(H_2A_kB) \\ &= P(H_1A_k)P_{H_1A_k}(B) + P(H_2A_k)P_{H_2A_k}(B) \\ &\to e^{-\Lambda u} \frac{(\Lambda u)^k}{k!} \cdot \frac{k!}{l_1!\, l_2! \dots l_m!} \left(\frac{u_1}{u}\right)^{l_1} \left(\frac{u_2}{u}\right)^{l_2} \dots \left(\frac{u_m}{u}\right)^{l_m} \\ &= e^{-\Lambda u} \Lambda^k \prod_{i=1}^{m} \frac{u_i^{l_i}}{l_i!}. \end{aligned}$$

Since the right-hand side of this relation coincides with the right-hand side of equation (16.2), our limit theorem is proved. Thus we can assert that, under the specified conditions that we have investigated, the total stream does in fact tend to a simple stream with parameter $\Lambda$.

# PART TWO

# SYSTEMS WITH LOSSES

## 17 Introductory remarks

Having, in the first part of this book, studied in detail the properties of a stream of incoming calls, we shall in the following parts investigate the fundamental questions connected with the serving of this stream. Every station (a point into which calls enter) is supplied with several apparatuses appointed to serve these calls. These apparatuses can be very varied according to the functions they are to fulfil; in particular even a person (a telephone operator, a shop assistant, a doctor) can act as such an "apparatus". For the sake of a unified terminology all serving apparatuses will be called "lines". The process of service operates in such a way that each incoming call occupies during a certain time one of the lines which is free (unoccupied) at the moment of its entry. While a line is busy with a call it is inaccessible to other incoming calls. The period of occupation of a line by one call will be called (again only for the sake of a unified terminology) a "conversation".

In problems of service, the difference between two types of organisation of stations is of fundamental significance. If there are free lines when a call enters in any organisation, the call occupies one of them and enters into a conversation. A divergence arises only when an incoming call finds all lines occupied. In the one organisation (systems with losses) such a call simply gets a refusal (or is "lost") and the subsequent course of the process of service continues as though this call had not occurred at all. In the other kind of organisation (systems allowing waiting) a call finding all lines occupied remains as an aspirant to a future conversation and thereafter occupies one of the lines which has become free. These two types of organisation differ from each other not only in details of the solution of basic problems but in their very structure. The fact is that the indices of the quality of service in these two instances are completely different. For a system with losses, the basic index is clearly the probability of a refusal (loss), a concept void of sense for the system which allows waiting. On the other hand for a system with waiting the central problem is the investigation of the "waiting time" as a random variable. Obviously, this problem has no meaning for a system with losses.

In view of the differences which exist, we should investigate these two types of structure completely separately. We shall devote the second part of the book to systems with losses and the third part to systems with waiting. Here we observe merely that even systems of an intermediate nature are credible. Thus, for example, it is possible that an unlucky call might have to wait provided that the number of waiting calls does not exceed a fixed limit after which any new incoming call gets a refusal. The theory of such mixed systems, which undoubtedly have a practical significance, has not yet been worked out.

Besides the main difference just indicated, the structures of the stations differ in many other features, in consequence of which the number of required structures differing among each other is very large. It is understandable that in our short monograph it will only be possible to present a small number of such structures and that therefore in solving each problem we have been obliged to start by accepting some given hypotheses which in reality are not the only possible ones. We will find it useful also to look at some further important features by which various systems serving calls may differ from one another.

(1) We shall suppose that all existing lines are always equally accessible to all incoming calls. (Such a system is called a "fully-accessible" collection of lines.) By contrast with this, it often happens in reality that some defined types of calls are "fixed" to a definite line and cannot occupy other lines.

(2) We shall nearly always assume a stream of incoming calls to be simple. In Part I it was indicated that this assumption corresponded closely to reality.

(3) We shall call a fully-accessible collection "ordered" if its lines are numbered in such a way that an incoming call always occupies a line with the smallest number out of those which are free at the moment of its entry (i.e. the first, if it is free; the second, if the first is occupied and the second free, etc.). In an "unordered" collection the lines are occupied at random. In practice, both types of organisation are to be found. It is as well to note that for a series of basic problems the question of orderedness, or otherwise, of a collection does not play any role; such, for example, is Erlang's problem, to which Chapters 6 and 7 will be devoted. However, problems do exist which, on the other hand, make sense only for an ordered collection. Such, for example, is Palm's problem which will be investigated in Chapter 8.

(4) For systems with waiting, the question of the order of service of waiting calls is of real significance in many problems. This service can be conducted either in the form of a stream or at random, and a whole series of problems may produce different results for each instance.

(5) Finally, the question of the distribution function of the length of occupation (conversations) is of very great significance for the majority of problems in the theory of service. In the great majority of investigations, this function is assumed to be exponential (i.e. the probability that the length of a conversation will be greater than $t$ is equal to $e^{-\beta t}$ where $\beta > 0$ is constant). This choice is conditioned mainly by the fact that it greatly facilitates necessary calculations. It may be said without exaggeration that most problems in the theory of service are solved with comparative ease by using an index of the distribution

of length of the conversation, and that, on the other hand, with almost any other assumption about the form of this law of distribution, they lead to insurmountable difficulties.

The central part played by the exponential distribution of the length of conversation is mainly due to one of its important properties which we shall use frequently hereafter. Suppose that $f_a(t)$ is the probability that a conversation which has lasted for $a$ seconds is prolonged for not less than $t$ seconds, so that for the exponential distribution $f_0(t) = e^{-\beta t}$. Since it is always true that

$$f_0(a+t) = f_0(a) f_a(t),$$

with the exponential distribution

$$e^{-\beta(a+t)} = e^{-\beta a} f_a(t),$$

from which

$$f_a(t) = e^{-\beta t}.$$

This means that *for the exponential distribution of the length of conversations the law of distribution of the remaining parts of a conversation does not depend upon its "increase", i.e. on however much time it has already lasted.* It is precisely this property of the exponential distribution that simplifies in most instances the calculations to be completed. It also leads one to think that in practical situations the hypothesis of an exponential distribution of the length of conversations can hardly be fully accurate and at best is only a reasonable approximation to reality.

CHAPTER 6

# ERLANG'S PROBLEM FOR A FINITE COLLECTION

## 18 Statement of the problem

This chapter will deal with a fully-accessible collection (whether ordered or not) of $n$ lines being entered by a simple stream of calls with parameter $\lambda$. We shall assume that the length of conversations is subject to the exponential law of distribution $1 - e^{-x}$. Since, for the general exponential law $1 - e^{-\beta x}$ the mean length of a conversation equals $1/\beta$, the choice $\beta = 1$ denotes simply that we are taking this mean length of the conversation as a unit of time. This, of course, in no way limits the general nature of the investigation.

If it is known that at a certain moment 0 precisely $k$ lines of a given collection ($0 \leqslant k \leqslant n$) are occupied, the number $N(t)$ of occupied lines at any following moment $t$ is a random variable the value of which is defined by a series of random factors — the moments of termination of those $k$ conversations which were in process at the moment 0, the moments of the occurrence of new calls between 0 and $t$, and the lengths of the conversations introduced by those calls. Thus the number $N(t)$ represents a one-parameter family of random variables or, as it is called, a *random process*. On the assumptions that we have made, this process has one important property which allows us to bring to the study of it methods which are well tried.

Let $N(t_0) = i$, i.e. at the moment $t_0$ there are $i$ lines occupied. Then the subsequent course of the process as far as probability is concerned does not depend on anything that has gone before the moment $t_0$. In fact, as already noted, this subsequent course is uniquely defined by the following three factors —

(1) The moments of termination of those conversations occurring at the moment $t_0$.

(2) The moments of the occurrence of new calls after the moment $t_0$.

(3) The lengths of the conversations for the calls mentioned in (2).

But it is easy to see that none of these three random factors depends on what occurred up to the moment $t_0$. For factor (1), this follows since, as seen in § 17, acceptance of an exponential distribution for the length of conversations implies that the length of the remaining part of a conversation does not depend on its previous history. For factor (2), it follows since the incoming stream of calls is simple and therefore does not have after-effects. Finally, for factor (3), it is self-evident. Thus all three factors mentioned are independent of what has gone before in our system, i.e. of the course of the process up to the moment

$t_0$. Consequently, the course of the process after the moment $t_0$ does not depend on what has gone before either, for it is uniquely defined by the three factors indicated.

Thus the random process $N(t)$ has the following property — if $N(t_0)$ is known, the course of the process after the moment $t_0$ as regards probability does not depend on its course up to the moment $t_0$ (briefly — if the present is known, the future does not depend on the past). Random processes having this property are called "Markoff processes".

If at a certain moment $t$, $i$ lines of a collection are occupied (i.e. $N(t) = i$), we shall say that at that moment the system is in "state $i$" ($i = 0, 1, 2, \dots n$). Thus, for the system $n + 1$ different states are possible. We shall denote by $P_{ik}(t)$ ($t > 0$, $0 \leqslant i \leqslant n$, $0 \leqslant k \leqslant n$) the conditional probability that a system being in the state $i$ at a certain moment after the passage of $t$ units changes to state $k$ (the probability $N(a+t) = k$ on condition $N(a) = i$). These "transition" probabilities play a fundamental role in the investigation of Markoff processes. Clearly

$$P_{ik}(t) \geqslant 0, \quad \sum_{k=0}^{n} P_{ik}(t) = 1 \qquad (0 \leqslant i \leqslant n,\ 0 \leqslant k \leqslant n).$$

If $t_1 > 0$, $t_2 > 0$, $0 \leqslant i \leqslant n$, $0 \leqslant k \leqslant n$, the following relation exists —

$$P_{ik}(t_1+t_2) = \sum_{r=0}^{n} P_{ir}(t_1) P_{rk}(t_2). \tag{18.1}$$

In fact, in order to go from state $i$ to state $k$ in time $t_1 + t_2$ the system should first change during time $t_1$ from state $i$ to some state $r$ and then during time $t_2$ change from state $r$ to state $k$, so that the relation (18.1) is the result of a simple change of the formula of compound probability. It is important to note that this formula is true only for Markoff processes. In fact, the transition probability $P_{rk}(t_2)$ can be reckoned independent of $i$ only for Markoff processes. If the process were not a Markoff process, then $P_{rk}(t_2)$ would need to be replaced by the probability of the change during time $t_2$ from state $r$ to state $k$, *assuming that the system during time $t_1$ changed preliminarily from state $i$ to state $r$*. Since for Markoff processes the probability $P_{rk}(t_2)$ does not depend on what occurred before this change, formula (18.1) remains true.

Formula (18.1), sometimes called the Chapman–Kolmogorov equation, lies at the basis of all investigations into Markoff processes. It will also play an important role in our exposition.

If at the starting moment 0, a system is in a given state $i$, the probability of finding it in state $k$ at any moment $t > 0$ is equal to $P_{ik}(t)$. However, we can make an assumption of a more general character regarding the starting moment — at the instant 0 we can regard as known not the state of the system but only the "first probability" $P_i(0)$ ($0 \leqslant i \leqslant n$) of various states. This general instance of course reduces itself to the above-mentioned particular instance when out of the numbers $P_i(0)$ any one is equal to unity (and the rest equal to nil).

The probability $P_k(t)$ of finding the system in the state $k$ at the moment $t$ is given by the formula of compound probability —

$$P_k(t) = \sum_{i=0}^{n} P_i(0)P_{ik}(t). \tag{18.2}$$

This probability depends as much upon the initial data $P_i(0)$ $(0 \leqslant i \leqslant n)$ as upon $t$.

If both sides of equation (18.1) are multiplied by $P_i(0)$ and summed for $i$ from 0 to $n$, then, by virtue of (18.2),

$$P_k(t_1+t_2) = \sum_{r=0}^{n} P_r(t_1)P_{rk}(t_2) \qquad (0 \leqslant k \leqslant n). \tag{18.3}$$

Erlang's problem, to which the present chapter is devoted, consists in evaluating the probability of finding a system in one or another given state. In the light of the above, such a statement of the question requires explanation. The random process $N(t)$ is non-stationary, the probabilities

$$P\{N(t) = k\} = P_k(t) \qquad (0 \leqslant k \leqslant n)$$

change in the course of time, and further depend on the preliminary data, i.e. on the numbers $P_i(0)$ $(0 \leqslant i \leqslant n)$. Thus it seems that the probabilities $P_k(t)$ sought in Erlang's problem can be defined only for given $t$ and $P_i(0)$ $(0 \leqslant i \leqslant n)$. In application, however, it is usually regarded possible to speak of the probability $p_k$ of finding a system in the state $k$, independent of the chosen moment of time and of the preliminary data. To justify such a practice theoretically, one can attempt to establish that as $t \to \infty$ the process $N(t)$ approximates without limit to a certain stationary process independent of the preliminary data. Speaking more concretely, it is necessary to establish that as $t \to \infty$ the probabilities $P_k(t)$ tend to certain constant numbers $p_k$ $(0 \leqslant k \leqslant n)$ independent of the first data. These figures $p_k$ may then be taken as the probabilities sought in Erlang's problem for finding a system in one or other defined state, since, on the one hand, the quantity $p_k$ does not depend on the preliminary data of the problem, and, on the other, comes as near as is needed to the real probability $P_k(t)$ if the process continues for a sufficiently long time.

Thus our task is to show that *as* $t \to \infty$ *the functions* $P_k(t)$ *tend to the quantities* $p_k$ $(0 \leqslant k \leqslant n)$ which do not depend on the preliminary data. Of course, doing this should provide values for the quantities $p_k$. In practice, the quantity $p_n$ is of particular significance; it is the probability of finding all lines occupied. This is the probability of "loss" (refusal), which in a system with losses is an important indicator of the quality of the service.

First of all, to reduce our task to one more suitable for the application of Chapman–Kolmogorov's equation, we shall use the following auxiliary proposition.

*Lemma. In order that as $t \to \infty$ the probabilities $P_k(t)$ should tend to the quantity $p_k$ $(0 \leqslant k \leqslant n)$ independent of the preliminary data, it is necessary and sufficient that the corresponding transition probabilities $P_{ik}(t)$ $(0 \leqslant k \leqslant n)$ should tend to the same limit for any value of $i$.*

*Proof.* Both assertions of the lemma are almost self-evident by virtue of (18.2).

(1) Suppose that $P_k(t) \to p_k$ $(t \to \infty, 0 \leqslant k \leqslant n)$, where $p_k$ does not depend on the preliminary data. Then by choosing $P_i(0) = 1$, $P_l(0) = 0$ $(l \neq i)$, we get $P_k(t) = P_{ik}(t)$ by virtue of (18.2), and consequently $P_{ik}(t) \to p_k$ $(t \to \infty,\ 0 \leqslant i \leqslant n,\ 0 \leqslant k \leqslant n)$.

(2) Suppose, on the other hand, $P_{ik}(t) \to p_k$ $(t \to \infty, 0 \leqslant i \leqslant n, 0 \leqslant k \leqslant n)$. Then, by virtue of (18.2), for any choice of probabilities $P_i(0)$ we get

$$P_k(t) \to \sum_{i=0}^{n} P_i(0) p_k = p_k \qquad (t \to \infty,\ 0 \leqslant k \leqslant n),$$

since $\sum_{i=0}^{n} P_i(0) = 1$.

By virtue of this lemma, our next problem is reduced to the proof that, as $t \to \infty$, the transition probabilities $P_{ik}(t)$ $(0 \leqslant i \leqslant n,\ 0 \leqslant k \leqslant n)$ tend to the limit $p_k$ independently of $i$.

## 19 Markoff's theorem

It would be possible to try to solve the above problem by finding expressions for the transition probabilities $P_{ik}(t)$ $(0 \leqslant k \leqslant n)$ for the special process $N(t)$ which interests us, and, by subsequently attempting an analysis of these expressions, to establish the existence and values of the limits. However, an alternative approach avoiding the difficult task of finding functions $P_{ik}(t)$ is to be preferred. In the present paragraph, without attempting to find the transition probabilities $P_{ik}(t)$ and their limits, we shall establish merely that these limits exist. This is possible because the existence theorem is a property of a very large class of Markoff processes and is by no means a special characteristic of our process $N(t)$.

Having done this, it will be possible to find these limits for the special process in which we are interested, again avoiding the explicit expressions of the functions $P_{ik}(t)$.

Let us agree to call Markoff's process, characterised by the transition probabilities $P_{ik}(t)$ $(0 \leqslant i \leqslant n,\ 0 \leqslant k \leqslant n)$, "transitive" if there exists such a $t > 0$ that $P_{ik}(t) > 0$ $(0 \leqslant i \leqslant n,\ 0 \leqslant k \leqslant n)$. Thus, for a transitive process there exists such a period of time in the course of which the change of a system from any state to any other is possible. It is immediately clear that the process $N(t)$ in which we are interested is transitive, and so any positive number can be selected in the capacity of $t$.

*Markoff's theorem*(*). *For any transitive Markoff process* $P_{ik}(t)$ $(0 \leqslant i \leqslant n, 0 \leqslant k \leqslant n)$ *the limit*

$$\lim_{t\to\infty} P_{ik}(t) = p_k \qquad (0 \leqslant i \leqslant n, 0 \leqslant k \leqslant n)$$

*exists and does not depend on i.*

*Proof.* Hereafter $k$ will denote an arbitrary fixed number of the series 0, 1, 2, . . . $n$. Suppose that

$$\max_{0\leq i\leq n} P_{ik}(t) = M_k(t), \quad \min_{0\leq i\leq n} P_{ik}(t) = m_k(t).$$

Thus, by virtue of (18.1), for any $i$ $(0 \leqslant i \leqslant n)$ and for any $t > 0, \tau > 0$

$$P_{ik}(t+\tau) = \sum_{r=0}^{n} P_{ir}(\tau)P_{rk}(t) \leqslant M_k(t) \sum_{r=0}^{n} P_{ir}(\tau) = M_k(t),$$

and consequently

$$M_k(t+\tau) \leqslant M_k(t),$$

i.e. $M_k(t)$ is a non-increasing function of $t$. Similarly, it is easy to establish that $m_k(t)$ is a non-decreasing function of $t$. It follows from this that as $t \to \infty$, $M_k(t)$ and $m_k(t)$ tend to fixed limits. Clearly the theorem will be proved if we show that these limits coincide, and to do this it is necessary and sufficient to have

$$\Delta_k(t) = M_k(t) - m_k(t) \to 0 \qquad (t \to \infty).$$

Hereafter, all sums for all indices will be extended to the values 0, 1, . . . , $n$ for these indices in consequence of which we shall not show the range of summation. Let $P_{ir}(t_0) > 0$ $(0 \leqslant i \leqslant n, 0 \leqslant r \leqslant n)$ (such a $t_0$ exists by virtue of the transitivity of the process). Suppose that

$$d_{il}^{(r)} = P_{ir}(t_0) - P_{lr}(t_0) \qquad (0 \leqslant i, l, r \leqslant n)$$

and hereafter denote by $\sum'$ (and correspondingly $\sum''$) sums extended only over the range of positive (and correspondingly non-positive) $d_{il}^{(r)}$. Then, by virtue of

$$\sum_r P_{ir}(t_0) = \sum_r P_{lr}(t_0) = 1 \qquad (0 \leqslant i, l \leqslant n),$$

it follows that

$$0 = \sum_r d_{il}^{(r)} = \sum_r{}' \left| d_{il}^{(r)} \right| - \sum_r{}'' \left| d_{il}^{(r)} \right| \qquad (0 \leqslant i, l \leqslant n)$$

or

$$\sum_r{}' \left| d_{il}^{(r)} \right| = \sum_r{}'' \left| d_{il}^{(r)} \right| = h_{il} \qquad (0 \leqslant i, l \leqslant n).$$

In this, by virtue of $P_{lr}(t_0) > 0$ $(0 \leqslant l, r \leqslant n)$,

(*)Markoff proved it for a "chain", i.e. processes with a discrete time; however, the proof can be transferred without change to the situation which interests us, of continuous time.

$$h_{il} = \sum_r{}' d_{il}^{(r)} = \sum_r{}'[P_{ir}(t_0) - P_{lr}(t_0)] < \sum_r{}' P_{ir}(t_0)$$
$$\leqslant \sum_r P_{ir}(t_0) = 1.$$

This inequality is true for any $i$ and $l$, in consequence of which

$$h = \max_{0 \leq i,\, l \leq n} h_{il} < 1.$$

Now let $q$ be any positive integer. Then, when $0 \leqslant i \leqslant n$, $0 \leqslant l \leqslant n$, from (18.1),

$$\begin{aligned}
&P_{ik}(qt_0+t_0) - P_{lk}(qt_0+t_0) \\
&= \sum_r P_{ir}(t_0)P_{rk}(qt_0) - \sum_r P_{lr}(t_0)P_{rk}(qt_0) \\
&= \sum_r [P_{ir}(t_0) - P_{lr}(t_0)]\, P_{rk}(qt_0) \\
&= \sum_r d_{il}^{(r)} P_{rk}(qt_0) \\
&= \sum_r{}' d_{il}^{(r)} P_{rk}(qt_0) - \sum_r{}'' \left| d_{il}^{(r)} \right| P_{rk}(qt_0) \\
&\leqslant M_k(qt_0)h_{il} - m_k(qt_0)h_{il} \\
&= h_{il}\Delta_k(qt_0) \leqslant h\Delta_k(qt_0).
\end{aligned}$$

Since this inequality is true for any $i$, $l$, it follows that we can take

$$P_{ik}(qt_0+t_0) = M_k(qt_0+t_0),$$
$$P_{lk}(qt_0+t_0) = m_k(qt_0+t_0),$$

and

$$\Delta_k(qt_0+t_0) \leqslant h\Delta_k(qt_0).$$

The recurrent application of this inequality gives

$$\Delta_k(qt_0) \leqslant h^{q-1}\Delta_k(t_0) \leqslant h^{q-1} \to 0 \qquad (q \to \infty).$$

By virtue of the monotonicity of the function $\Delta_k(t)$ it follows from this relation that

$$\Delta_k(t) \to 0 \qquad (t \to \infty).$$

Thus Markoff's theorem is proved.

## 20 Erlang's formulae and equations

We now turn to a classical method due to Erlang for deriving the magnitudes $p_k$ whose existence has just been proved. In contrast with the previous section, only the process $N(t)$ will here be dealt with.

In what follows a period of time of infinitesimally small length $\tau$ will be considered. For the sake of brevity $o(\tau)$ will be used to denote all infinitesimal quantities small by comparison with $\tau$, and the sign $\approx$ will be used to join any two expressions whose difference is a magnitude of the order $o(\tau)$.

By the assumption which was made in § 18, the probability $w(\tau)$ of the occurrence of at least one call in the period $\tau$ is a magnitude

$\approx \lambda\tau$, and the probability $\psi(\tau)$ of the occurrence of more than one call is a magnitude of the order $o(\tau)$. On the other hand, if at a given instant any line is occupied, the probability of its still being occupied for $\tau$ seconds (or more) is equal to $e^{-\tau}$. If $k$ lines are occupied, the probability that they will all remain occupied during the period $\tau$ is therefore equal to $e^{-k\tau}$. The probability that in course of time $\tau$ at least one of these lines will become free is

$$1 - e^{-k\tau} \approx k\tau.$$

The occurrence of calls and the freeing of lines provide elementary events at the moments of which the magnitude $N(t)$ changes discontinuously. From what has been said previously about the probabilities of such elementary events, it follows that the probability of the occurrence of at least one elementary event (of one or another kind) in a period of length $\tau$, is, as $\tau \to 0$, asymptotically proportional to $\tau$. Further, the probability of the occurrence in a period of this length of two or more elementary events (of whatever type) is a magnitude of order $o(\tau)$ (or, what amounts to the same, $\approx 0$).

These comments indicate how to find asymptotic expressions of the transition probabilities $P_{ik}(\tau)$ as $\tau \to 0$. First of all, if $|i - k| > 1$, the change from state $i$ to state $k$ clearly requires the occurrence of at least two elementary events. Thus from the above, as $\tau \to 0$,

$$P_{ik}(\tau) \approx 0 \qquad (|i - k| > 1). \tag{20.1}$$

Further, for the change from state $k < n$ to state $k + 1$, either the occurrence of one call or the occurrence of more than one elementary event is required. Therefore, from the above, as $\tau \to 0$,

$$P_{k,k+1}(\tau) \approx \lambda\tau \qquad (0 \leqslant k < n).$$

For a system to change from state $k > 0$ to state $k - 1$ either one of the lines must be freed or there must be an occurrence of more than one elementary event. Since the probability of one of the occupied lines $k$ being freed, over the time $\tau$, $\approx k\tau$ as $\tau \to 0$, it follows that

$$P_{k,k-1}(\tau) \approx k\tau \qquad (0 < k \leqslant n).$$

Finally, by virtue of (20.1), as $\tau \to 0$,

$$P_{kk}(\tau) \approx 1 - P_{k,k+1}(\tau) - P_{k,k-1}(\tau) \qquad (0 \leqslant k \leqslant n),$$

it being necessary to change the second and third terms of the right-hand side to zero when $k = n$ and $k = 0$, respectively. This gives

$$P_{00}(\tau) \approx 1 - \lambda\tau;$$
$$P_{kk}(\tau) \approx 1 - \lambda\tau - k\tau \qquad (1 \leqslant k \leqslant n - 1);$$
$$P_{nn}(\tau) \approx 1 - n\tau.$$

Thus, for all probabilities $P_{ik}(\tau)$, very simple asymptotic expressions giving them accurately to magnitudes of order $o(\tau)$ have been established.

Now consider equation (18.3) by virtue of which for any constant $t \geqslant 0$

$$P_k(t+\tau) = \sum_r P_r(t)P_{rk}(\tau).$$

Employing the above asymptotic expressions for the probabilities $P_{rk}(\tau)$ on the right-hand side of this equation, we get

$$\begin{aligned} P_0(t+\tau) &= P_0(t)P_{00}(\tau) + P_1(t)P_{10}(\tau) + o(\tau) \\ &= (1-\lambda\tau)P_0(t) + \tau P_1(t) + o(\tau); \\ P_k(t+\tau) &= P_{k-1}(t)P_{k-1,k}(\tau) + P_k(t)P_{kk}(\tau) + \\ &\qquad P_{k+1}(t)P_{k+1,k}(\tau) + o(\tau) \\ &= \lambda\tau P_{k-1}(t) + (1-\lambda\tau-k\tau)P_k(t) + \\ &\qquad (k+1)\tau P_{k+1}(t) + o(\tau) \qquad (0 < k < n); \\ P_n(t+\tau) &= P_{n-1}(t)P_{n-1,n}(\tau) + P_n(t)P_{nn}(\tau) + o(\tau) \\ &= \lambda\tau P_{n-1}(t) + (1-n\tau)P_n(t) + o(\tau); \end{aligned}$$

whence

$$\begin{aligned} \frac{P_0(t+\tau) - P_0(t)}{\tau} &= -\lambda P_0(t) + P_1(t) + o(1); \\ \frac{P_k(t+\tau) - P_k(t)}{\tau} &= \lambda P_{k-1}(t) - (\lambda+k)P_k(t) + \\ &\qquad (k+1)P_{k+1}(t) + o(1) \quad (0 < k < n); \\ \frac{P_n(t+\tau) - P_n(t)}{\tau} &= \lambda P_{n-1}(t) - nP_n(t) + o(1). \end{aligned}$$

If $\tau$ now tends to zero (preserving $t$ as a constant) the existence of all derivatives $P'_k(t)$ $(k = 0,1, \ldots, n)$ is established and, in the limit,

$$\left.\begin{aligned} P'_0(t) &= -\lambda P_0(t) + P_1(t) \\ P'_k(t) &= \lambda P_{k-1}(t) - (\lambda+k)P_k(t) + (k+1)P_{k+1}(t) \\ &\qquad\qquad (0 < k < n) \\ P'_n(t) &= \lambda P_{n-1}(t) - nP_n(t). \end{aligned}\right\} (\mathscr{E})$$

This system of $n + 1$ equations $(\mathscr{E})$ with $n + 1$ unknown functions $P_k(t)$ $(k = 0,1, \ldots, n)$ is called an *Erlang system*. Since all equations of this system are of the same type, the functions which we are seeking contain an arbitrary constant multiplier which may be defined using the "normalising" condition

$$\sum_{k=0}^{n} P_k(t) = 1.$$

As stated above, it is unnecessary to find solutions for the system of differential equations $(\mathscr{E})$. In § 19 it was shown that for any $k$ $(0 \leqslant k \leqslant n)$, the following limit exists —

$$\lim_{t\to\infty} P_k(t) = p_k.$$

It follows that as $t \to \infty$ the right-hand sides of all equations $(\mathscr{E})$ have limits. Going to the left-hand sides, it may be seen that as

$t \to \infty$ all derivatives $P_k'(t)$ tend to limits, but each of these limits can only be zero since, if any $P_k'(t)$ tended to other than zero, then, as $t \to \infty$, $|P_k(t)|$ would increase infinitely which — regardless of the real meaning of the magnitude $P_k(t)$ as a probability — is impossible on account of Markoff's theorem. Thus it may be concluded that

$$P_k'(t) \to 0 \quad (t \to \infty) \qquad (0 \leqslant k \leqslant n).$$

Consequently, the system $(\mathscr{E})$ in the limit as $t \to \infty$ gives

$$\left.\begin{aligned} -\lambda p_0 + p_1 &= 0 \\ \lambda p_{k-1} - (\lambda+k)p_k + (k+1)\,p_{k+1} &= 0 \qquad (0 < k < n) \\ \lambda p_{n-1} - np_n &= 0. \end{aligned}\right\} \quad (20.2)$$

This simple system of linear equations, together with the normalising condition $\sum_{k=0}^{n} p_k = 1$, uniquely determines the required numbers $p_k$.

If we set

$$\lambda p_{k-1} - kp_k = z_k \qquad (1 \leqslant k \leqslant n),$$

the system (20.2) can be written in the form

$$z_1 = 0, \quad z_k - z_{k+1} = 0 \qquad (0 < k < n), \quad z_n = 0;$$

whence $z_k = 0$ $(1 \leqslant k \leqslant n)$, and

$$p_k = \frac{\lambda}{k} p_{k-1} \qquad (1 \leqslant k \leqslant n)$$

and consequently

$$p_k = \frac{\lambda^k}{k!} p_0 \qquad (1 \leqslant k \leqslant n).$$

Substituting the normalising condition, it follows that

$$p_0 = \frac{1}{\sum\limits_{l=0}^{n} \dfrac{\lambda^l}{l!}}$$

and consequently that

$$p_k = \frac{\dfrac{\lambda^k}{k!}}{\sum\limits_{l=0}^{n} \dfrac{\lambda^l}{l!}} \qquad (k = 0, 1, \ldots, n). \qquad (20.3)$$

The formulae (20.3), usually called *Erlang's formulae*, fully solve the problem in hand. In particular, the probability of "loss" is given by the formula

$$p_n = \frac{\dfrac{\lambda^n}{n!}}{\sum\limits_{l=0}^{n} \dfrac{\lambda^l}{l!}}.$$

It is useful to note the decisive part that the assumption about the exponential distribution of the length of conversation has played in the whole of this investigation. Only on this assumption did the process $N(t)$ emerge as a Markoff process. If this assumption could not have been made, all the methods developed in §§ 19 and 20 would have been theoretically inapplicable. In specialist literature, there is a whole series of attempts to show that Erlang's formulae are valid even for other distributions of the length of conversations. However, as far as we can see, these attempts have not yet led to any conclusive results.

## 21 The ergodic theorem

Any probability in any theory only has a real meaning when a real collection of objects is known for which this probability gives the proportion of one kind or another. How real are those probabilities $p_k$ with which we were concerned in the last paragraphs? In particular, what does "the probability of loss", $p_n$, mean?

In the overwhelming majority of specialist investigations, the real interpretation of these probabilities has one and the same fully-defined character — they may be interpreted as "mean relative times of sojourn" of the system in corresponding states. What follows will show this. Let us denote by $x_k(t)$ the magnitude equal to 1, if a system at the instant $t$ is in state $k$ and equal to 0 otherwise. Then the integral

$$\int_0^T x_k(t)dt$$

represents the total length of those periods of time (between 0 and $T$) in the course of which the system is in state $k$. Further, the relation

$$\frac{1}{T}\int_0^T x_k(t)dt$$

gives the "mean relative time of sojourn" of the system in state $k$ (during the period $(0,T)$). In the limit, the probability $p_k$ of finding the system in state $k$ is thus

$$\lim_{T\to\infty} \frac{1}{T}\int_0^T x_k(t)dt. \tag{21.1}$$

However, it is clear that the magnitudes $p_k$ defined in the preceding paragraphs do not immediately permit of a similar interpretation. This can be seen from the fact that the magnitude $x_k(t)$ represents from our point of view a random function (random process), and consequently that the limit (21.1) (if it exists) is a random variable which cannot therefore be identified with the probability $p_k$ since the former depends on the particular instance. On the other hand, the probabilities $p_k$ were defined as limits as $t \to \infty$ of the probabilities $P_k(t)$. If all probabilities are taken as mean times of sojourn of a system in one or another state, then the magnitudes $P_k(t)$ do not permit of any reasonable interpretation as one can clearly see.

Thus our definition of probabilities $p_k$, and also our method of calculating them, do not immediately provide a basis for identifying them with limits of the kind (21.1), despite the practice established in all applied literature. If this identification is impossible from the practical point of view, it is highly desirable to devise means by which the integrals

$$X_k(T) = \frac{1}{T}\int_0^T x_k(t)dt$$

may be calculated for large values of $T$, if only by showing their approximate coincidence with the probabilities $p_k$. If this could be done, then such an approach would largely justify the generally accepted practice in specialist literature of taking the probabilities $p_k$ as "mean relative times of sojourn," a practice very convenient in applied problems.

Since $X_k(T)$ is from our point of view a random variable, its approximation to the magnitude $p_k$, regardless of the particular instance, can at best be asserted only with a certain (sufficiently high) probability. In the present chapter, we shall show that these speculations are correct. That is, the following theorem is true.

*Theorem. For any* $\varepsilon > 0$, *however small,*

$$\lim_{T\to\infty} P\{|X_k(T) - p_k| > \varepsilon\} = 0$$

*Preliminary remarks.* In theoretical physics, propositions which show that the probability of a system being in a certain state (defined as a part in a great collection of systems of identical state) tends over a long period of time in any one system to the mean time of sojourn in the given state are called *ergodic theorems.* The proposition to be proved represents a typical example of an ergodic theorem.

*Proof.* In what follows, the preliminary data (the probabilities $P_k(0)$, $0 \leqslant k \leqslant n$) are arbitrary, but fixed. The mathematical expectation of the random variable $\xi$ will be denoted by $M\xi$ or $M\{\xi\}$. Since the variable $x_k(t)$ can bear the value 1 and 0 with corresponding probabilities $P_k(t)$ and $1 - P_k(t)$, it follows that

$$Mx_k(t) = P_k(t). \tag{21.2}$$

Since $P_k(t) \to p_k$ as $t \to \infty$, it also follows that

$$\int_0^T [P_k(t) - p_k]\, dt = o(T) \qquad (T \to \infty).$$

Thus, for sufficiently large $T$

$$P\{|X_k(T) - p_k| > \varepsilon\} = P\left\{\left|\int_0^T [x_k(t) - p_k]\, dt\right| > \varepsilon T\right\}$$

$$< P\left\{\left|\int_0^T [x_k(t) - P_k(t)]\, dt\right| > \frac{\varepsilon}{2} T\right\}$$

Substituting the inequality of Chebishev, we find by virtue of (21.2),

$$P\{|X_k(T) - p_k| > \varepsilon\} < \frac{4}{\varepsilon^2 T^2} M\left\{\left[\int_0^T [x_k(t) - P_k(t)]\, dt\right]^2\right\}$$

$$= \frac{4}{\varepsilon^2 T^2} M\left\{\int_0^T \int_0^T [x_k(u) - P_k(u)]\,[x_k(v) - P_k(v)]\, du\, dv\right\}$$

$$= \frac{4}{\varepsilon^2 T^2} \int_0^T \int_0^T [M\{x_k(u)x_k(v)\} - P_k(u)P_k(v)]\, du\, dv$$

$$= \frac{8}{\varepsilon^2 T^2} \int\int_{0 \leq u \leq v \leq T} [M\{x_k(u)x_k(v)\} - P_k(u)P_k(v)]\, du\, dv.$$

Since the magnitude $x_k(u)x_k(v)$ can take only the values 1 and 0, $M\{x_k(u)x_k(v)\} = P\{x_k(u)x_k(v) = 1\}$ is the probability of finding the system in the same state $k$ at the instant $u$ as at the instant $v$. Thus, for $v > u$,

$$M\{x_k(u)x_k(v)\} = P_k(u)P_{kk}(v-u)$$

and we find

$$P\{|X_k(T) - p_k| > \varepsilon\}$$

$$< \frac{8}{\varepsilon^2 T^2} \int\int_{0 \leq u \leq v \leq T} P_k(u)\,[P_{kk}(v-u) - P_k(v)]\, du\, dv$$

$$= \frac{8}{\varepsilon^2 T^2} \int_0^T P_k(u) du \int_u^T [P_{kk}(v-u) - P_k(v)]\, dv$$

$$= \frac{8}{\varepsilon^2 T^2} \int_0^T P_k(u) du \int_0^{T-u} [P_{kk}(z) - P_k(z+u)]\, dz.$$

Suppose now that $\delta > 0$ is arbitrarily small and $A$ is so large that for $z > A$,

$$|P_{ik}(z) - p_k| < \frac{\delta}{2} \qquad (0 \leqslant i \leqslant n).$$

Then, for $z > A$, $u > 0$,

$$|P_{kk}(z) - p_k| < \frac{\delta}{2},$$

$$\begin{aligned} |P_k(z+u) - p_k| &= |\sum_i P_i(0) P_{ik}(z+u) - p_k| \\ &= |\sum_i P_i(0)\, [P_{ik}(z+u) - p_k]| \\ &< \frac{\delta}{2} \sum_i P_i(0) = \frac{\delta}{2}, \end{aligned}$$

and consequently

$$|P_{kk}(z) - P_k(z+u)| < \delta.$$

Thus, for $T > A$,

$$\begin{aligned} P\{|X_k(T) - p_k| > \varepsilon\} &\leqslant \frac{8}{\varepsilon^2 T^2} \int_0^T du \int_0^T |P_{kk}(z) - P_k(z+u)|\, dz \\ &\leqslant \frac{8}{\varepsilon^2 T^2} \int_0^T du \left\{ \int_0^A dz + \int_A^T \delta\, dz \right\} \\ &\leqslant \frac{8}{\varepsilon^2 T^2} \left\{ AT + \delta T^2 \right\} = \frac{8A}{\varepsilon^2 T} + \frac{8\delta}{\varepsilon^2}. \end{aligned}$$

By first taking a sufficiently small $\delta$, by then selecting $A$ in the manner described above, and finally, by selecting $T$ sufficiently large, it may be seen that the right-hand side of the last inequality is arbitrarily small for sufficiently large $T$. This proves the theorem.

CHAPTER 7

# ERLANG'S PROBLEM FOR AN INFINITE COLLECTION

## 22 Equations for the generating function

If the number of lines in a collection is infinite the above type of system should no longer be included among "systems with losses" since losses now become impossible. However, the calculation of the probabilities of various states retains practical significance even in this instance, as situations occur in practice where losses are inadmissible and the number of lines must be sufficiently large for the probability of losses to be negligible. In such instances, the probabilities of various states provide means whereby the degree of use of the system may be evaluated, and this in its turn is required for the calculation of the speed of effecting the service and other economic indices

If, in Erlang's formulae, $n \to \infty$, then

$$p_k = e^{-\lambda} \frac{\lambda^k}{k!} \qquad (k = 0,1,2, \ldots).$$

Thus it is possible to foresee that this is a Poisson distribution and gives us the probabilities of various states for an infinite collection. However, the method by which we changed formulae (20.3) is unacceptable for an infinite collection, since Markoff's theorem, on which it is founded, makes the essential assumption that the number of states is finite. It will now be shown that the case of an infinite collection can be studied quite simply by the method of generating functions.

The arguments employed in § 20 to derive Erlang's system ($\mathscr{E}$) of equations are almost completely valid for an infinite collection. The only change is that the group of equations of the system ($\mathscr{E}$), which we constructed in § 20 for $0 < k < n$, now holds for any $k > 0$, and the last of the system's equations is completely eliminated. Thus, for probabilities of different states for an infinite collection, we derive the system of equations

$$\left.\begin{aligned} P_0'(t) &= -\lambda P_0(t) + P_1(t) \\ P_k'(t) &= \lambda P_{k-1}(t) - (\lambda+k)\, P_k(t) + (k+1)\, P_{k+1}(t) \qquad (k > 0). \end{aligned}\right\} (\mathscr{E}^*)$$

This system is simpler than system ($\mathscr{E}$), since, here, for all $k > 0$ all equations are of an identical type. This circumstance allows the method of generating functions to be applied to the solution of system ($\mathscr{E}^*$). (It cannot be applied directly to system ($\mathscr{E}$).)

Suppose that

$$\Phi(t,x) = \sum_{k=0}^{\infty} P_k(t)x^k$$

(this series converges for $|x| \leqslant 1$ and for any $t$). Suppose again that for $k > 0$

$$\lambda P_{k-1}(t) - kP_k(t) = Q_k(t).$$

Then the system of equations ($\mathscr{E}^*$) achieves the short formulation —

$$P_0'(t) = - Q_1(t)$$
$$P_k'(t) = Q_k(t) - Q_{k+1}(t) \qquad (k > 0).$$

Thus

$$\begin{aligned} \frac{\partial \Phi}{\partial t} &= \sum_{k=0}^{\infty} P_k'(t)x^k = - Q_1(t) + \sum_{k=1}^{\infty} [Q_k(t) - Q_{k+1}(t)]x^k \\ &= (x-1) \sum_{k=1}^{\infty} Q_k(t)x^{k-1} \\ &= (x-1)\{\lambda \sum_{k=1}^{\infty} P_{k-1}(t)x^{k-1} - \sum_{k=1}^{\infty} kP_k(t)x^{k-1}\} \\ &= (x-1)\left\{\lambda\Phi - \frac{\partial \Phi}{\partial x}\right\} \end{aligned}$$

or

$$\frac{\partial \Phi}{\partial t} + (x-1)\frac{\partial \Phi}{\partial x} - \lambda(x-1)\Phi = 0. \tag{22.1}$$

This simple partial differential equation of the first order serves to determine the function $\Phi(t,x)$. First of all, the equation may be further simplified by transforming the unknown function. Suppose that

$$\exp [\lambda(x-1)(1-e^{-t})] = G(t,x)$$

and that

$$\Phi(t,x) = G(t,x)F(t,x),$$

where $F(t,x)$ is the new unknown function. Then

$$\frac{\partial \Phi}{\partial t} = G\frac{\partial F}{\partial t} + FG\lambda(x-1)e^{-t};$$

$$\frac{\partial \Phi}{\partial x} = G\frac{\partial F}{\partial x} + FG\lambda(1-e^{-t}),$$

from which

$$\begin{aligned} \frac{\partial \Phi}{\partial t} + (x-1)\frac{\partial \Phi}{x} - \lambda(x-1)\Phi = G\Big\{\frac{\partial F}{\partial t} + \lambda(x-1)e^{-t}F + \\ (x-1)\frac{\partial F}{\partial x} + \lambda(x-1)(1-e^{-t})F - \lambda(x-1)F\Big\} \\ = G\left\{\frac{\partial F}{\partial t} + (x-1)\frac{\partial F}{\partial x}\right\}, \end{aligned}$$

and hence equation (22.1) is equivalent to the equation

$$\frac{\partial F}{\partial t} + (x-1)\frac{\partial F}{\partial x} = 0. \tag{22.2}$$

Now, suppose that

$$(x-1)e^{-t} = L(t,x).$$

Then the Jacobian

$$\frac{\partial(F,L)}{\partial(t,x)} = e^{-t}\left\{\frac{\partial F}{\partial t} + (x-1)\frac{\partial F}{\partial x}\right\}$$

and hence equation (22.2) is equivalent to the equation

$$\frac{\partial(F,L)}{\partial(t,x)} = 0.$$

A general solution to this is

$$F = R(L) = R[(x-1)e^{-t}]$$

where $R$ is an arbitrary differentiable function of its argument. The required function $\Phi$ is then given by the expression

$$\Phi(t,x) = \exp\left[\lambda(x-1)(1-e^{-t})\right] R[(x-1)e^{-t}]. \tag{22.3}$$

This is the general solution of equation (22.1). For the solution of the problem the form of function $R$ must now be determined using the preliminary data.

## 23 Solution of the problem

Choose the first moment $t = 0$. Then

$$P_k(0) = a_k \qquad (k = 0,1\ldots).$$

In particular, if $t = 0$ and the system is in state $i$, $a_i = 1$, $a_k = 0$ $(k \neq i)$. In all instances $\sum_{k=0}^{\infty} a_k = 1$. Then, setting $t = 0$ in the expression (22.3) of the function $\Phi(t,x)$ gives

$$\Phi(0,x) = \sum_{k=0}^{\infty} P_k(0)x^k = \sum_{k=0}^{\infty} a_k x^k = R(x-1).$$

Hence

$$R(z) = \sum_{k=0}^{\infty} a_k(z+1)^k,$$

and

$$R[(x-1)e^{-t}] = \sum_{k=0}^{\infty} a_k[1 + (x-1)e^{-t}]^k, \tag{23.1}$$

and the expression for the generating function is

$$\Phi(t,x) = \exp\,[\lambda(x-1)\,(1-e^{-t})] \sum_{k=0}^{\infty} a_k[1 + (x-1)e^{-t}]^k \text{(*)}$$

The problem is, by using this expression, to find the limiting values of the probabilities $P_k(t)$ as $t \to \infty$.

As the relation (23.1) shows, the expression $R[(x-1)e^{-t}]$ can be presented as a power series in $x$ (converging when $t > 0$, $|x| \leqslant 1$) —

$$R[(x-1)e^{-t}] = \sum_{k=0}^{\infty} b_k(t)x^k.$$

Let us establish now that as $t \to \infty$

$$b_0(t) \to 1, \quad b_k(t) \to 0 \qquad (k > 0).$$

In fact, by virtue of (23.1),

$$\sum_{k=0}^{\infty} b_k(t)x^k = R[(x-1)e^{-t}] = \sum_{k=0}^{\infty} a_k(xe^{-t} + 1 - e^{-t})^k. \qquad (23.2)$$

Since $a_k \geqslant 0$, $e^{-t} > 0$ and $1 - e^{-t} \geqslant 0$, comparison of the left-hand side with the right shows, first, that all $b_k(t) \geqslant 0$. Furthermore, setting $x = 0$ in this equation gives

$$b_0(t) = \sum_{k=0}^{\infty} a_k(1-e^{-t})^k$$

from which, as $t \to \infty$,

$$b_0(t) \to \sum_{k=0}^{\infty} a_k = 1.$$

Finally, setting $x = 1$ in (23.2), it follows that for any $t \geqslant 0$

$$\sum_{k=0}^{\infty} b_k(t) = \sum_{k=0}^{\infty} a_k = 1,$$

and from this, by virtue of the above, $b_k(t) \to 0$ ($k > 0$, $t \to \infty$) and our assertion is proved.

Now, by virtue of (22.3), it follows that

$$\Phi(t,x) = \sum_{n=0}^{\infty} P_n(t)x^n = R[(x-1)e^{-t}] \exp\,[\lambda(x-1)\,(1-e^{-t})]$$

$$= \sum_{k=0}^{\infty} b_k(t)x^k \exp\,[-\,\lambda(1-e^{-t})] \sum_{r=0}^{\infty} \frac{\lambda^r(1-e^{-t})^r}{r!}\, x^r,$$

from which

$$P_n(t) = \exp\,[-\,\lambda(1-e^{-t})] \sum_{k=0}^{n} b_k(t)\, \frac{\lambda^{n-k}(1-e^{-t})^{n-k}}{(n-k)!} \qquad (n \geqslant 0)$$

(*) This series converges directly for $t > 0$, $|x| \leqslant 1$. In fact it is also easy to establish that under these conditions $|1 + (x-1)e^{-t}| \leqslant 1$.

Now if, for a fixed $n$, $t$ is increased indefinitely, then, from the above, $b_0(t) \to 1$, $b_k(t) \to 0$ $(k > 0)$ and so

$$P_n(t) \to e^{-\lambda} \frac{\lambda^n}{n!} \qquad (t \to \infty, n = 0,1,2\ldots)$$

which completely solves the problem in hand. The same formulae are thus obtained as were derived at the beginning of § 22 as a limiting form of Erlang's formula for a finite collection.

## 24 Stream with a variable parameter

The methods advanced in §§ 22 and 23 provide a simple means of solving Erlang's problem for the infinite collection, even when the parameter of an incoming stream of calls varies with time. This situation was encountered in § 5. Now and hereafter $\lambda(t)$ will be used to denote the "instantaneous value" at the moment $t$ of the parameter $\lambda$ defined in (5.1).

As the equations of the system $(\mathscr{E}^*)$ in § 22 often have a local nature (corresponding to some defined moment of time $t$) the deduction made from these equations still holds for the situation where the parameter varies, but replacing the constant $\lambda$ there is now the "instantaneous value" $\lambda(t)$ of this parameter, generally speaking different at different moments of time $t$. Thus, as an initial system of equations for the definition of functions $P_k(t)$ we have

$$\left.\begin{aligned} P_0'(t) &= -\lambda(t)P_0(t) + P_1(t) \\ P_k'(t) &= \lambda(t)P_{k-1}(t) - [\lambda(t) + k]\, P_k(t) + (k+1)\, P_{k+1}(t) \\ &\qquad\qquad\qquad\qquad\qquad\qquad (k > 0). \end{aligned}\right\} \qquad (24.1)$$

Assuming that

$$\Phi(t,x) = \sum_{k=0}^{\infty} P_k(t)x^k,$$

then from the system (24.1), exactly as in § 22, the generating function $\Phi(t,x)$ may be shown to satisfy the partial differential equation

$$\frac{\partial \Phi}{\partial t} + (x-1)\frac{\partial \Phi}{\partial x} - \lambda(t)\,(x-1)\Phi = 0 \qquad (24.2)$$

which differs from equation (22.1) only in that the constant parameter $\lambda$ is replaced by the function $\lambda(t)$. This difference, not essential for deriving equation (24.2), affects its solution and necessitates an alternative substitution for the function required.

In what follows assume that

$$e^{-t} \int_0^t e^u \lambda(u) du = \Lambda(t);$$

$$\exp\,[(x-1)\Lambda(t)] = G(t,x);$$

$$\Phi(t,x) = G(t,x)F(t,x),$$

where $F(t,x)$ is the new unknown function. Substituting this in the equation (24.2) the function $F(t,x)$ satisfies the equation

$$\frac{\partial F}{\partial t} + (x-1)\frac{\partial F}{\partial x} = 0,$$

which is the same as equation (22.2). In § 22, it was shown that a general solution of this equation is

$$F(t,x) = R[(x-1)e^{-t}],$$

where $R$ is an arbitrary differentiable function of its argument. Thus

$$\Phi(t,x) = \exp\,[(x-1)\Lambda(t)]\; R[(x-1)e^{-t}]. \tag{24.3}$$

Now the form of the function $R$ may now be determined using the preliminary data $P_k(0) = a_k$ $(k = 0,1,2,\ldots)$. As in § 23, this is

$$R(z) = \sum_{k=0}^{\infty} a_k(z+1)^k$$

whence, in particular,

$$R[(x-1)e^{-t}] = \sum_{k=0}^{\infty} a_k[1 + (x-1)e^{-t}]^k \tag{24.4}$$

where the series converges for $t \geqslant 0$, $|x| \leqslant 1$. From this, using (24.3), the final expression of the generating function is derived as

$$\Phi(t,x) = \exp\,[(x-1)\Lambda(t)] \sum_{k=0}^{\infty} a_k[1 + (x-1)e^{-t}]^k. \tag{24.5}$$

In § 23, it was assumed that

$$R[(x-1)e^{-t}] = \sum_{k=0}^{\infty} b_k(t)x^k, \tag{24.6}$$

and established that, as $t \to \infty$,

$$b_0(t) \to 1, \quad b_k(t) \to 0, \qquad (k > 0).$$

These deductions hold true in the present instance also, since the new function $R(z)$ is the same as previously.

By virtue of (24.5), (24.4) and (24.6), it follows that

$$\Phi(t,x) = \sum_{n=0}^{\infty} P_n(t)x^n = \exp\,[(x-1)\Lambda(t)] \sum_{k=0}^{\infty} b_k(t)x^k$$

$$= e^{-\Lambda(t)} \sum_{r=0}^{\infty} \frac{[\Lambda(t)]^r}{r!}\, x^r \sum_{k=0}^{\infty} b_k(t)x^k$$

$$= e^{-\Lambda(t)} \sum_{n=0}^{\infty} \left\{ \sum_{r=0}^{n} \frac{[\Lambda(t)]^r}{r!}\, b_{n-r}(t) \right\} x^n$$

and consequently that

$$P_n(t) = e^{-\Lambda(t)} \sum_{r=0}^{n} \frac{[\Lambda(t)]^r}{r!}\, b_{n-r}(t) \qquad (n = 0,1,2,\ldots).$$

Now suppose that as $t \to \infty$ the parameter $\lambda(t)$ remains bounded;

$$\lambda(t) < \Lambda \qquad (t \geqslant 0).$$

Then it is also true that

$$\Lambda(t) = e^{-t} \int_0^t e^u \lambda(u) du < \Lambda \qquad (t \geqslant 0).$$

Thus, for any constant $n$ as $t \to \infty$, all terms in the summation

$$\sum_{r=0}^{n} \frac{[\Lambda(t)]^r}{r!} b_{n-r}(t)$$

except the last term ($r = n$) tend to zero, while the last differs infinitesimally from $[\Lambda(t)]^n/n!$ Thus, as $t \to \infty$ and for constant $n$,

$$P_n(t) - e^{-\Lambda(t)} \frac{[\Lambda(t)]^n}{n!} \to 0$$

i.e. the law of distribution $P_n(t)$ approaches Poisson's law with a (variable) parameter

$$\Lambda(t) = e^{-t} \int_0^t e^u \lambda(u) du.$$

This completes the solution.

## 25 The infinite collection with an arbitrary distribution function for the lengths of conversation

In §§ 22–24 it was established that for Erlang's problem an infinite collection is more simply investigated than the finite one. (In particular, § 24 completed the investigation of the case of an incoming stream with a variable parameter, a problem which, as far as we know, has not yet been solved for a finite collection.) In our presentation, the comparative simplicity of the treatment of the case of an infinite collection has arisen as a result of employing the method of generating functions. This, for the sake of showing the close analogy with the theory of a finite collection, has always started from Erlang's system of differential equations. However, Erlang's problem for the infinite collection is a simple problem which can be solved by quite elementary methods. If these are employed, it is found that the exponential distribution function of the length of conversations, an important assumption in Erlang's method of equations, can without undue complication be replaced by any other law. With this important observation, we shall now proceed with an elementary deduction of Erlang's formulae for an infinite collection.

All the propositions of § 22 will be retained except that the probability $F(x)$ of a conversation chosen at random having a length $> x$ will now be assumed to be an arbitrary non-increasing function subject only to the requirements

$$F(0) = 1,\ F(+\infty) = 0,\ -\int_0^\infty x dF(x) = \int_0^\infty F(x)dx = 1,$$

in which the last chooses the unit of time to be the mean length of the conversation. As before, denote by $P_k(t)$ the probability that at any moment $t > 0, k$ conversations occur (or, what amounts to the same, that $k$ lines are occupied). Our problem is to show that

$$\lim_{t\to\infty} P_k(t) = e^{-\lambda}\frac{\lambda^k}{k!} \qquad (k = 0,1,2,\ldots).$$

Suppose that $X = (x_1, x_2, \ldots, x_r)$ is an arbitrary $r$-dimensional vector, belonging to the region $D_r$ $(0 < x_1 < x_2 < \ldots < x_r < t)$ of an $r$-dimensional space. Let us call "hypothesis $(r,X)$" the situation where, during the period $(0,t)$, there occur $r$ calls, and the moments $t_i$ $(0 < t_1 < t_2 < \ldots < t_r < t)$ of these calls satisfy the inequalities

$$x_i < t_i < x_i + dx_i \qquad (1 \leqslant i \leqslant r).$$

Since the stream of calls is of simple type with parameter $\lambda$, the probability of hypothesis $(r,X)$ is accurately expressed for small $dx_i$ by the formula

$$P(r,X) = \lambda e^{-\lambda x_1} dx_1 \lambda \exp[-\lambda(x_2 - x_1)]dx_2 \ldots$$
$$\lambda \exp[-\lambda(x_r - x_{r-1})]dx_r \exp[-\lambda(t - x_r)]$$

where the last term is the probability that between moments $x_r$ and $t$ no calls occur. From this

$$P(r,X) = e^{-\lambda t}\lambda^r dx_1\, dx_2 \ldots dx_r \tag{25.1}$$

(and, in particular, does not depend on the vector $X$). Suppose $P_{k,(r,X)}(t)$ is the conditional probability of finding $k$ conversations at the moment $t$, if hypothesis $(r,X)$ occurs. Clearly, $P_{k,(r,X)}(t)$ can differ from zero only if $r \geqslant k$. Since, for a conversation beginning at the instant $x$ $(0 < x < t)$, the probability of its not ending till the instant $t$ is equal to $F(t-x)$, thus

$$P_{k,(r,X)}(t) = \sum_C \prod_{i=1}^{k} F(t - x_{s_i}) \prod_{\substack{s=1\\ s\neq s_i}}^{r} [1 - F(t - x_s)], \tag{25.2}$$

where $C$ is any combination $(s_1, s_2, \ldots, s_k)$ of $k$ numbers from the series $(1,2,\ldots,r)$ and where the summation occurs over all combinations.

By the formula of compound probability,

$$P_k(t) = \sum_{r,X} P(r,X)P_{k,(r,X)}(t),$$

where $r$ is summed from $k$ to $\infty$ and $X$ is integrated over the region $D_r$. Hence, by virtue of (25.1) and (25.2)

$$P_k(t) = e^{-\lambda t}\sum_{r=k}^{\infty}\lambda^r \int_{D_r} \sum_C \prod_{i=1}^{k} F(t - x_{s_i}) \prod_{\substack{s=1\\ s\neq s_i}}^{r} [1 - F(t - x_s)]\, dx_1 \ldots dx_r.$$

Here, the function under the integral sign is symmetric with regard to the variables $x_1, x_2 \dots, x_r$. Thus the integral does not change if in the defined region $D_r$ the series $x_1 < x_2 < \dots < x_r$ of variables of integration is replaced by any other series. But there are $r!$ such series, and since the combination of all regions $D_r$ obtained in this way gives the $r$-dimensional cube $K_r$ $0 < x_i < t$, $i = 1,2, \dots, r$, the integration may be carried out over this cube if we divide by $r!$. Thus

$$P_k(t) = e^{-\lambda t} \sum_{r=k}^{\infty} \frac{\lambda^r}{r!} \sum_C \int_{K_r} \prod_{i=1}^{k} F(t-x_{s_i}) \prod_{\substack{s=1 \\ s \neq si}}^{r} [1-F(t-x_s)] \, dx_1 \dots dx_r.$$

Here, the integral over the cube $K_r$ divides into $r$ simple integrals and is equal to

$$\left\{ \int_0^t F(t-u)du \right\}^k \left\{ \int_0^t [1 - F(t-u)]du \right\}^{r-k}$$
$$= \left\{ \int_0^t F(z)dz \right\}^k \left\{ \int_0^t [1 - F(z)]dz \right\}^{r-k}.$$

Assuming that

$$\int_t^{\infty} F(z)dz = \varepsilon(t),$$

it follows that

$$\int_0^t F(z)dz = 1 - \varepsilon(t), \qquad \int_0^t [1 - F(z)]dz = t - 1 + \varepsilon(t),$$

and that — since the number of combinations $C$ is equal to $\binom{r}{k}$ —

$$P_k(t) = e^{-\lambda t} \sum_{r=k}^{\infty} \frac{\lambda^r}{r!} \binom{r}{k} [1 - \varepsilon(t)]^k [t - 1 + \varepsilon(t)]^{r-k}$$
$$= e^{-\lambda t} \frac{\lambda^k}{k!} [1 - \varepsilon(t)]^k \sum_{r=k}^{\infty} \frac{\lambda^{r-k}}{(r-k)!} [t - 1 + \varepsilon(t)]^{r-k}$$
$$= e^{-\lambda t} \frac{\lambda^k}{k!} [1 - \varepsilon(t)]^k \exp \lambda [t - 1 + \varepsilon(t)]$$
$$= e^{-\lambda} \frac{\lambda^k}{k!} [1-\varepsilon(t)]^k \exp \lambda\varepsilon(t).$$

Since $\varepsilon(t) \to 0$ as $t \to \infty$, it follows that

$$\lim_{t\to\infty} P_k(t) = e^{-\lambda} \frac{\lambda^k}{k!} \qquad (k = 0,1,2 \dots),$$

which was to be proved(*).

---

(*)In this proof it was not clearly stated that the stream of calls began at the moment 0, i.e. $P_0(0) = 1$, $P_k(0) = 0$ $(k > 0)$. The proof would be only a little more difficult for arbitrary preliminary data.

CHAPTER 8

# PALM'S PROBLEM

## 26 Statement of the problem

In solving Erlang's problem in Chapters 6 and 7, it did not matter whether a given collection of lines was regarded as ordered or not. In this chapter, however, we shall consider a problem which makes sense only for an ordered collection. Thus it will always be assumed that the lines of a given collection are numbered and that each incoming call occupies the line with the lowest number amongst those that are free at the moment of its entry. Clearly, in such an arrangement the mean business will be different for different lines — the busiest will be the first, after that the second, and so on.

On the other hand, the problems investigated in this chapter will be seen not to depend upon whether the collection is finite or infinite. Thus the difference which was important for Erlang's problem is irrelevant in the present chapter.

In what follows the assumptions of the previous chapters will be retained. The incoming stream of calls is assumed to be simple with parameter $\lambda$. The lengths of conversations are assumed to be independent of one another and of any data about the occurrence of calls, and to be distributed according to the exponential law with unit mean.

We shall denote by $L_r$ the line with the number $r$. For all the arguments of this chapter, the simple and self-evident fact that the combination of lines $L_1, L_2, \ldots, L_r$ (for any $r$ not exceeding the general number of lines in the collection) can be regarded as an independent, fully-accessible collection is of fundamental importance. Each call "lost" in this collection (i.e. finding the first $r$ of the lines occupied) enters the line $L_{r+1}$ (if, of course, there is one), and conversely for a call to enter $L_{r+1}$ it is essential for it to have been lost to the collection ($L_1$, $L_2, \ldots, L_r$). The probability of loss in this collection is the proportion of time during which all lines $L_1, L_2, \ldots, L_r$ are occupied. It is obviously quite independent of whether there are lines with higher numbers and how many of them there are. It can be calculated by Erlang's formula for the probability of loss on a collection of $r$ lines and is equal to

$$\frac{\dfrac{\lambda^r}{r!}}{\displaystyle\sum_{i=0}^{r} \frac{\lambda^i}{i!}}.$$

In particular, for $r = 1$, the probability of the loss of a call on $L_1$ is given by the expression

$$\frac{\lambda}{1+\lambda} \tag{26.1}$$

and, for $r = 2$, the probability of loss from the collection $(L_1, L_2)$ is

$$\frac{\frac{1}{2}\lambda^2}{1+\lambda+\frac{\lambda^2}{2}} = \frac{\lambda^2}{\lambda^2+2\lambda+2}. \tag{26.2}$$

But the loss of a call on collection $(L_1, L_2)$ may be regarded as a double event — (1) a loss to $L_1$, and (2) a loss to $L_2$. The probability of the first of these instances is equal to $\lambda/(1+\lambda)$. To find the conditional probability of the second event, on condition that the first took place, we observe that it is the probability of loss on $L_2$ for a call lost to $L_1$ (or, equivalently, for one entering $L_2$). If $\lambda'$ is the intensity of a stream of calls entering into $L_2$, the required conditional probability of the second event will, by the formula (26.1), be equal to $\lambda'/(1+\lambda')$ (as $L_2$ is the first line of this stream). But $\lambda'$ is the number of calls lost to $L_1$ in a unit of time. Since $\lambda$ calls occur in $L_1$ in a unit of time, and since the proportion of losses is $\lambda/(1+\lambda)$, then

$$\lambda' = \lambda\,\frac{\lambda}{1+\lambda} = \frac{\lambda^2}{1+\lambda},$$

and the required conditional probability of the second event is equal to

$$\frac{\lambda'}{1+\lambda'} = \frac{\lambda^2}{1+\lambda+\lambda^2}.$$

The probability of loss on collection $(L_1, L_2)$ is thus given by the expression

$$\frac{\lambda}{1+\lambda}\,\frac{\lambda^2}{1+\lambda+\lambda^2} = \frac{\lambda^3}{(1+\lambda)\,(1+\lambda+\lambda^2)}, \tag{26.3}$$

which differs from the expression (26.2) obtained directly from Erlang's formulae. Consequently, formula (26.3) for the probability of loss on collection $(L_1, L_2)$ is erroneous(*). By examining the chain of reasoning which led to this formula, the source of the error is soon revealed. Denoting by $\lambda'$ the intensity of a stream of calls entering into $L_2$ and employing Erlang's formula (26.1), the probability of loss to $L_2$ of a call entering this line was taken as $\lambda'/(1+\lambda')$. In this, it was wrongly assumed that the stream of calls entering into $L_2$ was "of simple type", since Erlang's formulae were established on this assumption. That an erroneous conclusion was reached shows that this assumption was invalid. Thus, in effect, it has been established that *the stream of calls entering into* $L_2$ (*or what is the same thing lost to* $L_1$) *is not a simple stream.* Furthermore, there is no reason to expect that streams of calls entering into $L_3$, $L_4$, . . . are simple ones. This negative deduction is

(*)I am obliged to V. K. Leserson for pointing out this "paradox".

instructive in that it emphasises how in the solution of the most elementary tasks it is important not to restrict investigation to simple incoming streams.

A detailed and thorough investigation of the nature of streams of calls entering any line $L_r$ of a given ordered fully-accessible collection is a task of theoretical interest, and is also of practical importance in the solution of the problem to which the present chapter is devoted. All the basic results in this field were obtained by Palm (8) in 1943.

## 27 Elementary calculations

As was shown in the previous section, the probability that a call entering into a collection (or, what amounts to the same thing, into line $L_1$) is lost to the collection $(L_1, L_2, \ldots, L_r)$ is, from Erlang's formula,

$$E_r = \frac{\dfrac{\lambda^r}{r!}}{\sum\limits_{k=0}^{r} \dfrac{\lambda^k}{k!}} \qquad (r = 1, 2, \ldots).$$

It is clear that the call is lost to line $L_r$ when and only when it is lost to the collection $(L_1, L_2, \ldots, L_r)$. Thus one can also say that the number $E_r$ expresses the probability of loss to line $L_r$. However, it is essential to remember that this refers to the probability of loss to $L_r$ *for a call entering into* $L_1$. The probability $\pi_r$ of loss to $L_r$ for a call entering into $L_r$ is a different quantity which will now be investigated. For this purpose, it will be assumed that $r > 1$ as $\pi_1 = E_1$.

The probability $E_r$ that a call which has come into $L_1$ will be lost to $L_r$ can be represented as the product of two factors — the probability $E_{r-1}$ that it will be lost to $L_{r-1}$ and the conditional probability of its loss to $L_r$ if it is known that it is lost to $L_{r-1}$ (i.e. has entered into $L_r$). But this conditional probability is also $\pi_r$. Thus

$$E_r = E_{r-1}\, \pi_r$$

or

$$\pi_r = \frac{E_r}{E_{r-1}} \qquad (r > 1).$$

Apart from its apparent simplicity, this formula is unsuitable for our problem in that it contains both $E_r$ and $E_{r-1}$. Thus, a more suitable formula is

$$\pi_r = \frac{\lambda}{r + \lambda E_{r-1}} \qquad (r > 1) \qquad (27.1)$$

which will now be proved. Assume for the sake of brevity that

$$\sum_{k=0}^{q} \frac{\lambda^k}{k!} = S_q \qquad (q \geqslant 0)$$

so that

$$E_r = \frac{1}{S_r} \frac{\lambda^r}{r!} \qquad (r \geqslant 1).$$

Consequently, when $r > 1$,

$$\frac{1}{\pi_r} = \frac{E_{r-1}}{E_r} = \frac{r}{\lambda}\frac{Sr}{S_{r-1}} = \frac{r}{\lambda}\left[1 + \frac{\lambda^r/r!}{S_{r-1}}\right] = \frac{1}{\lambda}(r + \lambda E_{r-1})$$

and this gives (27.1). It is necessary to note that $\lambda$ in the formula (27.1) denotes the intensity of a primitive stream of calls entering $L_1$.

It is of interest to compare the probabilities of loss to different lines with entering streams of equal intensities. The calculations produced in all cases show that the probability increases with the number of the line. In Palm's work, he asserts that this follows directly from formula (27.1). However, we do not see how this is so and it is doubtful whether this assertion in its general formulation is true. We have only managed to show the following, more modest proposition.

*Theorem. The probability of loss to $L_r$ when $r > 1$ is always greater than the probability of loss to $L_1$ if the streams entering into these lines are of identical intensity.*

*Proof.* If a stream entering into $L_1$ has intensity $\lambda$, then the probability of entering into $L_r$ having entered $L_1$ is equal to $E_{r-1}$ (since this is the probability of there being a loss on lines $L_1, L_2, \ldots, L_{r-1}$). Thus, on average, among $\lambda$ calls entering $L_1$ in a unit of time, $\lambda E_{r-1}$ will enter $L_r$, i.e. the stream of calls entering into $L_r$ will have an intensity $\lambda E_{r-1}$. From the above, the probability of loss to $L_r$ for calls of this stream is equal to

$$\pi_r = \frac{\lambda}{r + \lambda E_{r-1}} \qquad (r > 1).$$

If a stream of this same intensity, $\lambda E_{r-1}$, entered $L_1$, the probability of loss to $L_1$ for calls of this stream, from formula (26.1), would be

$$\frac{\lambda E_{r-1}}{1 + \lambda E_{r-1}} \qquad (r > 1).$$

For the proof of our theorem, it is therefore sufficient to establish that if $r > 1$ and $\lambda > 0$, then

$$\frac{\lambda E_{r-1}}{1 + \lambda E_{r-1}} < \frac{\lambda}{r + \lambda E_{r-1}},$$

or

$$r + \lambda E_{r-1} < \lambda + \frac{1}{E_{r-1}},$$

i.e.

$$\lambda E_{r-1} + r - \lambda < \frac{1}{E_{r-1}} \qquad (r > 1). \qquad (27.2)$$

Suppose that $r = 2$, $E_{r-1} = E_1 = \dfrac{\lambda}{1+\lambda}$ then (27.2) takes the form

$$\frac{\lambda^2}{1+\lambda} + 2 - \lambda < \frac{1+\lambda}{\lambda}$$

which is equivalent to

$$\frac{\lambda^2}{1+\lambda} + 1 - \lambda < \frac{1}{\lambda}$$

or

$$\frac{1}{1+\lambda} < \frac{1}{\lambda}$$

and consequently is fulfilled for any $\lambda > 0$.

Now suppose that the inequality (27.2) is true for any $r > 1$ (and for any $\lambda > 0$). It will be shown that in this instance (and even when $r$ is replaced by $r + 1$, i.e. for any $\lambda > 0$) the following inequality holds —

$$\lambda E_r + r + 1 - \lambda < \frac{1}{E_r}.$$

By virtue of (27.1) (and remembering that $\pi_r = E_r/E_{r-1}$),

$$E_r = \frac{\lambda E_{r-1}}{r + \lambda E_{r-1}}.$$

From this, it follows that

$$\lambda E_r + r + 1 - \lambda - \frac{1}{E_r} = r\left\{\frac{\lambda E_{r-1} + r - \lambda}{\lambda E_{r-1} + r} - \frac{1}{\lambda E_{r-1}}\right\}$$

$$= \frac{rH_r(\lambda)}{\lambda E_{r-1}(\lambda E_{r-1} + r)}$$

where

$$H_r(\lambda) = \lambda E_{r-1}(\lambda E_{r-1} + r - \lambda) - (\lambda E_{r-1} + r)$$

$$= \lambda E_{r-1}\left(\lambda E_{r-1} + r - \lambda - \frac{1}{E_{r-1}}\right) - (\lambda E_{r-1} + r - \lambda)$$

$$= (\lambda E_{r-1} - 1)(\lambda E_{r-1} + r - \lambda) - \lambda.$$

The theorem will be proved if it can be established that $H_r(\lambda) < 0$ for any $\lambda > 0$. Suppose that

$$\lambda E_{r-1} - 1 = A, \quad \lambda E_{r-1} + r - \lambda = B,$$

so that

$$H_r(\lambda) = AB - \lambda.$$

It may be assumed that $AB > 0$, since, if this were not true, obviously $H_r(\lambda) < 0$. If $A > 0$ and if $B > 0$ then, since $B < 1/E_{r-1}$ from (27.2), it follows that

$$AB < \frac{\lambda E_{r-1} - 1}{E_{r-1}} < \lambda.$$

Further, if $A < 0$, $B < 0$, then clearly

$$|A| < 1, \quad |B| = -B = \lambda - r - \lambda E_{r-1} < \lambda,$$
$$AB = |A|\,|B| < \lambda.$$

Thus $H_r(\lambda) < 0$ in all cases, and the theorem is proved.

## 28 Palm's fundamental theorem

In § 26 it was established that if a simple stream of calls enters line $L_1$ then the stream entering line $L_r$ ($r > 1$) cannot be simple. The most important theorem in Palm's theory is that when this is so *there enters into any line $L_r$ a stream of the type P, i.e. stationary, orderly, and with limited after-effects*. Consequently, such a stream differs from a simple one only in that the requirement of absence of after-effects is changed for the more general requirement of limited after-effects.

For the proof of this theorem, no further results are required. It is sufficient to look more closely at what has been said before, since it is only necessary to show that if it is true for $L_r$, it remains true also for $L_{r+1}$ ($r = 1,2, \ldots$). Further, since calls entering $L_{r+1}$ coincide with calls lost to $L_r$ it is necessary to prove the following — *if a stream of calls of type P enters the line $L_r$, then the calls lost to $L_r$ also form a stream of type P*. When the problem is stated in this way, the very existence of the line $L_{r+1}$ may be seen to be superfluous.

For brevity, denote by $A$ the stream of calls entering $L_r$, and by $B$ the stream of calls lost to $L_r$. The course of stream $B$ after any arbitrarily-chosen moment $t_0$ is uniquely defined if it is known how long a conversation occupying line $L_r$ at the moment $t_0$ will last, and at what moments after $t_0$ calls of the stream $A$ enter and what are the lengths of conversations of these calls. But all of these factors are uniquely defined by the course of stream $A$ which is a stream of the type $P$ and consequently stationary. Thus all the stated factors do not depend on the moment $t_0$ chosen, and together with them the subsequent course of the stream $B$ does not depend on $t_0$ either. In other words, stream $B$ is also stationary.

The orderliness of stream $B$ results from the fact that it represents a part of the (orderly) stream $A$.

Finally, it will be shown that stream $B$ has a limited after-effect. Let $t_0 = \tau_0 = 0$. Denote by $t_i$ ($i = 1,2, \ldots$), the moments of calls in stream $A$ after $t_0$, and by $\tau_1, \tau_2, \ldots$ the moments of calls in stream $B$ after $t_0$, and further suppose that

$$\tau_k - \tau_{k-1} = \xi_k \qquad (k = 1,2, \ldots).$$

Our problem is to show that the distribution function of the variable $\xi_{k+1}$ is independent of the variables $\xi_1, \xi_2, \ldots \xi_k$ or, alternatively, of the variables $\tau_1, \tau_2, \ldots \tau_k$. However, the variable $\xi_{k+1}$ is uniquely defined if the following are known —

(1) The distance from $\tau_k$ to further calls entering $L_r$ after the moment $\tau_k$.

(2) The subsequent length of the conversation occupying line $L_r$ at the moment $\tau_k$.

(3) The lengths of conversations occupying the line $L_r$ after the moment $\tau_k$.

All these three factors are independent of the known variables $\tau_1, \tau_2, \ldots \tau_k$. For the first, this follows from the fact that stream $A$ has limited after-effect. For the second, from the exponential law of the length of conversations and, for the third, it is self-evident. But, since the value of the random variable $\xi_{k+1}$ is uniquely defined by these three factors, it follows that this random variable does not depend on the variables $\tau_1, \tau_2, \ldots \tau_k$, nor, alternatively on the variables $\xi_1, \xi_2, \ldots \xi_k$. But this also implies that stream $B$ has a limited after-effect. Thus the basic theorem of Palm is proved.

It follows from this that for any ordered fully-accessible collection with a simple incoming stream and an exponential distribution for the lengths of conversations the stream of calls entering into line $L_r$ of this collection is a stream of type $P$. But such a stream (*vide* § 13) is uniquely defined by the Palm function $\varphi(t)$. We shall therefore denote by $\varphi_r(t)$ $(r = 0,1,2\ldots)$, Palm's function for the stream of calls entering line $L_r + 1$ (so that, in particular $\varphi_0(t) = e^{-\lambda t}$). It may be seen that the problem reduces to determining the function $\varphi_r(t)$ for any $r > 0$. What follows will be concerned with this.

### 29 Deduction from the basic system of equations

Suppose a call is lost to line $L_r$ at the moment $t_0$ (or, what amounts to the same thing, a call enters the line $L_{r+1}$). Then $\varphi_r(t)$ is the probability that in the period $(t_0, t_0 + t)$ no call will be lost to $L_r$ (i.e. enters $L_{r+1}$). But this event can arise in two ways —

($A$) In the period $(t_0, t_0 + t)$ no call enters $L_r$.

($B$) In the period $(t_0, t_0 + t)$, calls enter $L_r$, but none of them are lost.

Thus we have

$$\varphi_r(t) = P(A) + P(B),$$

in which, when $r > 0$, $P(A) = \varphi_{r-1}(t)$. Thus, on the one hand a call lost to $L_r$ at the moment $t_0$ will also be lost to $L_{r-1}$, but on the other hand the stream of calls entering $L_r$ will coincide with the stream of calls lost to $L_{r-1}$.

We now determine $P(B)$. Suppose that the first call entering $L_r$ after the moment $t_0$ occurs in the period $(t_0 + x,\ t_0 + x + dx)$. The probability of this occurrence is $\varphi_{r-1}(x) - \varphi_{r-1}(x+dx) = -d\varphi_{r-1}(x)$. For this call not to be lost to $L_r$ it is necessary and sufficient that after the moment $t_0 + x$ the line $L_r$ should be freed from the conversation with which it was occupied at the moment $t_0$. The probability of this is equal to $1 - e^{-x}$. Thus the probability that the first call after the moment $t_0$ will enter $L_r$ in the period $(t_0 + x,\ t_0 + x + dx)$ and that this call will not be lost, is equal to

$$-\ (1-e^{-x})d\varphi_{r-1}(x).$$

It can now be stated that if the above occurs and if $x < t$, the probability that no call will be lost to $L_r$ during the remaining period $(t_0 + x,\ t_0 + t)$ is equal to $\varphi_r(t-x)$. This would emerge directly from the definition of the function $\varphi_r(t)$ if the call entering $L_r$ in the period $(t_0 + x,\ t_0 + x + dx)$ were lost to this line. But in fact from our assumptions this call is not lost, so that this assertion requires proving. Whether the call entering $L_r$ at the moment $t_0 + x$ (or more exactly in the period $(t_0 + x,\ t_0 + x + dx)$), is lost or not, once this call has entered, the line $L_r$ is in any case occupied from the moment $t_0 + x$. When it is freed, by virtue of the exponential law of distribution of the lengths of conversations this in no way depends on whether it was occupied by the call entering at the moment $t_0 + x$ or whether it was occupied earlier (and the call entering at the moment $t_0 + x$ was lost). On the other hand, the moments of calls entering $L_r$ later (after $t_0 + x$) depend on the fact that such a call occurred at the moment $t_0 + x$, but do not depend on the service of this call (on whether it was lost or not). Clearly, the lengths of conversations of these subsequent calls do not depend on this service. Thus none of the factors defining the presence or absence of losses to $L_r$ in the period $(t_0 + x,\ t_0 + t)$ depends on the kind of service received by a call entering at the moment $t_0 + x$. Thus, although, according to our assumption, this call was not lost, the probability that in the period $(t_0 + x,\ t_0 + t)$ there will be no losses to the line $L_r$ will be the same as if it had been lost, i.e. equal to $\varphi_r(t-x)$. Combining this with what was stated earlier, the following conclusion is reached — when $x < t$, the probability that the first call after the moment $t_0$ will occur on $L_r$ in the period $(t_0 + x,\ t_0 + x + dx)$, and that between $t_0$ and $t_0 + t$ no call will be lost to $L_r$, is equal to

$$-\ (1-e^{-x})d\varphi_{r-1}(x)\varphi_r(t-x).$$

But to find the probability of occurrence $B$, all such probabilities must be summed for $x$ from 0 to $t$. This gives

$$P(B) = -\int_0^t (1-e^{-x})\varphi_r(t-x)d\varphi_{r-1}(x)$$

and consequently

$$\varphi_r(t) = \varphi_{r-1}(t) - \int_0^t (1-e^{-x})\varphi_r(t-x)d\varphi_{r-1}(x) \qquad (r \geqslant 1). \tag{29.1}$$

This is also the initial system of equations in Palm's theory. From (29.1), it is immediately clear that for any $t > 0$

$$\varphi_r(t) \geqslant \varphi_{r-1}(t),$$

an inequality which is self-evident.

## 30 Laplace's transformation

To solve the fundamental system of equations (29.1) the required functions $\varphi_r(t)$ will be subjected to Laplace transformations. Suppose that

$$\psi_r(t) = \int_0^\infty e^{-tx}\varphi_r(x)dx \qquad (r \geqslant 0)$$

and write the equation (29.1) in the form

$$\varphi_r(x) = \varphi_{r-1}(x) - \int_0^x (1-e^{-z})\varphi_r(x-z)d\varphi_{r-1}(z).$$

Now, multiply both sides by $e^{-tx}$, and integrate $x$ from 0 to $\infty$. This gives

$$\psi_r(t) = \psi_{r-1}(t) - \int_0^\infty e^{-tx}dx \int_0^x (1-e^{-z})\varphi_r(x-z)d\varphi_{r-1}(z)$$

$$= \psi_{r-1}(t) - \int_0^\infty (1-e^{-z})e^{-zt}d\varphi_{r-1}(z) \int_z^\infty e^{-t(x-z)}\varphi_r(x-z)dx$$

$$= \psi_{r-1}(t) - \int_0^\infty (1-e^{-z})e^{-zt}d\varphi_{r-1}(z) \int_0^\infty e^{-ty}\varphi_r(y)dy$$

$$= \psi_{r-1}(t) - \psi_r(t) \int_0^\infty (1-e^{-z})e^{-zt}d\varphi_{r-1}(z). \tag{30.1}$$

Integration by parts gives

$$\int_0^\infty e^{-zt}d\varphi_{r-1}(z) = -1 + t\psi_{r-1}(t)$$

and consequently

$$\int_0^\infty e^{-z(t+1)}d\varphi_{r-1}(z) = -1 + (t+1)\psi_{r-1}(t+1).$$

Thus from (30.1) we get

$$\psi_r(t) = \psi_{r-1}(t) - \psi_r(t)\,[t\psi_{r-1}(t) - (t+1)\psi_{r-1}(t+1)],$$

from which

$$\psi_r(t) = \frac{\psi_{r-1}(t)}{1 + t\psi_{r-1}(t) - (t+1)\psi_{r-1}(t+1)}. \tag{30.2}$$

Thus, for determining the functions $\psi_r(t)$ a simple recurrence formula is derived. Since $\varphi_0(x) = e^{-\lambda x}$, it follows that

$$\psi_0(t) = \int_0^\infty e^{-(\lambda+t)x} dx = \frac{1}{t+\lambda}$$

and the relation (30.2) thus allows us to determine all functions $\psi_r(t)$. In particular, these functions may directly be seen to be rational. However, this is not sufficient since to transform back to the function $\varphi_r(x)$ it is necessary to know more about the properties of these rational functions $\psi_r(t)$. In particular, for this reverse transformation, the factorisation of functions $\psi_r(t)$ into partial fractions is important and consequently the nature and distributions of the roots of their denominators. These problems will now be investigated.

It should be noted that by a simple transformation of the required functions,

$$t\psi_r(t) = \chi_r(t) \qquad (r = 0,1,2\ldots)$$

the recurrence formulae (30.2) may be reduced to a simpler form

$$\chi_r(t) = \frac{\chi_{r-1}(t)}{1 + \chi_{r-1}(t) - \chi_{r-1}(t+1)}.$$

However, this fact will not be used.

## 31 Determination of functions $\psi_r(t)$

Denote by $B_r(t)$ $(r = -1,0,1,2,\ldots)$, the polynomial of degree $r+1$

$$B_r(t) = \lambda^{r+1} + \sum_{l=0}^{r} \binom{r+1}{l} t\,(t+1)\ldots(t+r-l)\lambda^l,$$

where $\sum\limits_{l=0}^{-1} = 0$ and consequently $B_{-1}(t) = \lambda^0 = 1$.

*Lemma. The polynomials $B_r(t)$ are linked by the recurrence formula*

$$B_r(t) = tB_{r-1}(t+1) + \lambda B_{r-1}(t) \qquad (r = 0,1,\ldots). \quad (31.1)$$

*Proof.* We have

$$\lambda B_{r-1}(t) = \lambda^{r+1} + \sum_{l=0}^{r-1} \binom{r}{l} t(t+1)\ldots(t+r-1-l)\lambda^{l+1}$$

$$tB_{r-1}(t+1) = t\lambda^r + \sum_{l=0}^{r-1} \binom{r}{l} t(t+1)\ldots(t+r-l)\lambda^l$$

$$= \sum_{l=0}^{r} \binom{r}{l} t(t+1)\ldots(t+r-l)\lambda^l$$

Since $\binom{r}{0} = \binom{r+1}{0}$ and, for $l > 0$, $\binom{r}{l-1} + \binom{r}{l} = \binom{r+1}{l}$

we get

$$\lambda B_{r-1}(t) + tB_{r-1}(t+1) = \lambda^{r+1} + \sum_{l=1}^{r} \binom{r+1}{l} t(t+1)\ldots(t+r-l)\lambda^l + \binom{r}{0} t(t+1)\ldots(t+r)$$

$$= \lambda^{r+1} + \sum_{l=0}^{r} \binom{r+1}{l} t(t+1)\ldots(t+r-l)\lambda^l = B_r(t),$$

which proves the lemma.

Now a simple expression can be found for the rational functions $\psi_r(t)$.

*Theorem.*

$$\psi_r(t) = \frac{B_{r-1}(t+1)}{B_r(t)} \qquad (t = 0,1,2\ldots). \tag{31.2}$$

*Proof.* Since $B_{-1}(t) = 1$ and since it is easy to see that $B_0(t) = t + \lambda$, for $r = 0$ the relation (31.2) to be proved has the form

$$\psi_0(t) = \frac{1}{t+\lambda},$$

and was established in § 30. Thus we may suppose that $r > 0$ and that the relation (31.2) is already established for $\psi_{r-1}(t)$. That it remains true for $\psi_r(t)$ also will be shown. Our theorem is then proved by induction.

By virtue of this assumption, it follows that

$$1 + t\psi_{r-1}(t) - (t+1)\psi_{r-1}(t+1)$$

$$= 1 + \frac{tB_{r-2}(t+1)}{B_{r-1}(t)} - \frac{(t+1)B_{r-2}(t+2)}{B_{r-1}(t+1)}$$

$$= \frac{K_r(t)}{B_{r-1}(t)B_{r-1}(t+1)},$$

where

$$K_r(t) = B_{r-1}(t+1)B_{r-1}(t) + tB_{r-2}(t+1)B_{r-1}(t+1) - (t+1)B_{r-2}(t+2)B_{r-1}(t). \tag{31.3}$$

Thus, from (30.2),

$$\psi_r(t) = \frac{B_{r-2}(t+1)}{B_{r-1}(t)} \frac{B_{r-1}(t)B_{r-1}(t+1)}{K_r(t)}$$

$$= \frac{B_{r-2}(t+1)B_{r-1}(t+1)}{K_r(t)}.$$

The relation (31.2) to be proved is thus equivalent to

$$K_r(t) = B_r(t)B_{r-2}(t+1),$$

which in its turn from (31.1) is equivalent to

$$K_r(t) = tB_{r-1}(t+1)B_{r-2}(t+1) + \lambda B_{r-1}(t)B_{r-2}(t+1).$$

Writing expression (31.3) instead of $K_r(t)$ it is then necessary to prove the following relation,

$$B_{r-1}(t+1)B_{r-1}(t) - (t+1)B_{r-2}(t+2)B_{r-1}(t) = \lambda B_{r-1}(t)B_{r-2}(t+1),$$

or

$$B_{r-1}(t+1) = (t+1)B_{r-2}(t+2) + \lambda B_{r-2}(t+1).$$

But this relation is obtained directly by substituting $r - 1$ for $r$ and $t + 1$ for $t$ in the recurrence formula (31.1). Thus the theorem is proved.

It may be seen that each function $\psi_r(t)$ is a regular rational fraction whose numerator is a polynomial of degree $r$ and whose denominator is of degree $r + 1$.

## 32 Expansion of functions $\psi_r(t)$ in partial fractions

To determine the functions $\varphi_r(x)$ by their Laplace transformations $\psi_r(t)$, it is necessary to investigate how the rational functions $\psi_r(t)$ expand into partial fractions and for this purpose to determine the roots of the polynomials $B_r(t)$ which serve as the denominators of these rational functions.

The polynomial $B_r(t)$ of degree $r + 1$, as can be seen from its definition, has positive coefficients in which the coefficient for $t^{r+1}$ is equal to 1. From this it follows that all real roots of this polynomial are negative and that it can be written in the form

$$B_r(t) = (t+a_{r0})\,(t+a_{r1}) \ldots (t+a_{rr}),$$

where the numbers $a_{ri}$ $(0 \leqslant i \leqslant r)$ are positive or complex. It may now be asserted that *the numbers $a_{ri}$ are all positive and that if they are distributed in order of size, then*

$$a_{ri} > a_{r,i-1} + 1 \qquad (1 \leqslant i \leqslant r),$$

i.e. the distance between two neighbouring roots exceeds unity.

This assertion will be proved by induction. It has already been seen that $B_0(t) = t + \lambda$, so that the assertion is true for $r = 0$. Suppose that it is true for $B_r(t)$ $(r \geqslant 0)$ and it will now be seen that it then holds also for $B_{r+1}(t)$. Using the recurrence formula (31.1) for this gives

$$B_{r+1}(t) = tB_r(t+1) + \lambda B_r(t) = t(t+1+a_{r0}) \ldots (t+1+a_{rr}) + \lambda(t+a_{r0}) \ldots (t+a_{rr}).$$

Here, from our assumption, all $a_{rk} > 0$ and $a_{rk} > a_{r,k-1} + 1$ $(1 \leqslant k \leqslant r)$.

This formula (and also the indirect definition) shows that $B_{r+1}(0) > 0$. On the other hand, it gives

$$B_{r+1}(-a_{r0}) = -a_{r0}(a_{r1}+1-a_{r0}) \ldots (a_{rr}+1-a_{r0}) < 0.$$

This shows that $B_{r+1}(t)$ has a root between 0 and $-a_{r0}$.

Now, suppose that $k$ is one of the numbers of the series $0, 1, \ldots, r-1$. Then

$$B_{r+1}(-a_{rk}-1) = \lambda(a_{r0}-a_{rk}-1)\ldots(a_{r,k-1}-a_{rk}-1) \times (a_{rk}-a_{rk}-1)(a_{r,k+1}-a_{rk}-1)\ldots(a_{rr}-a_{rk}-1)$$

has, as it is easy to calculate, $k+1$ negative factors and consequently the sign $(-1)^{k+1}$. Against this,

$$B_{r+1}(-a_{r,k+1}) = -a_{r,k+1}(-a_{r,k+1}+1+a_{r0})\ldots(-a_{r,k+1}+1+a_{rk}) \times (-a_{r,k+1}+1+a_{r,k+1})\ldots(-a_{r,k+1}+1+a_{rr})$$

has $k+2$ negative factors and consequently the sign $(-1)^k$. Thus the polynomial $B_{r+1}(t)$ for any $k$ $(0 \leqslant k \leqslant r-1)$ has a root in the interval between $-a_{rk}-1$ and $-a_{r,k+1}$.

Finally, since the leading term of the polynomial $B_{r+1}(t)$ is $t^{r+2}$, for negative $t$ with sufficiently large modulus, $B_{r+1}(t)$ has the sign $(-1)^{r+2}$. At the same time, since in the expression

$$B_{r+1}(-a_{rr}-1) = \lambda(-a_{rr}-1+a_{r0})\ldots(-a_{rr}-1+a_{rr})$$

each bracket expression is negative, it has the sign $(-1)^{r+1}$. This shows that $B_{r+1}(t)$ has a root between $-a_{rr}-1$ and $-\infty$.

Combining all that has been said, it may be seen that the polynomial $B_{r+1}(t)$ has at least one root in each of the $r+2$ intervals —

$$(0, -a_{r0}),\ (-a_{rk}-1, -a_{r,k+1}) \quad (0 \leqslant k \leqslant r-1), \quad (-a_{rr}-1, -\infty)$$

which in pairs do not have common points. Since $B_{r+1}(t)$ is a polynomial of degree $r+2$, its roots are exhausted by this enumeration. Further, since the $r+2$ intervals under investigation are such that the distance between two neighbouring roots of the polynomial $B_{r+1}(t)$ always exceeds unity, our assertion about the distribution of roots of polynomials $B_r(t)$ is fully proved.

It follows from this that the expansion of the function $\psi_r(t)$ into partial fractions has the form

$$\psi_r(t) = \sum_{k=0}^{r} \frac{C_{rk}}{t+a_{rk}}$$

where the numerators $C_{rk}$ can easily be expressed in terms of the $a_{rk}$ by well-known methods.

## 33 Conclusion

Since the function $\dfrac{C_{rk}}{t+a_{rk}}$ is the Laplace transform of the function $C_{rk} \exp - a_{rk}x$, so the function $\psi_r(t)$, whose form has just been found, is the Laplace transform of the function

$$\sum_{k=0}^{r} C_{rk} \exp - a_{rk}x. \tag{33.1}$$

But $\psi_r(t)$ was defined as the Laplace transform of the required function

$\varphi_r(x)$. Can it be concluded from this that $\varphi_r(x)$ is given by the function (33.1)?

The theory of Laplace transformations (the details of which cannot be discussed here) shows that of the functions satisfying a given transformation of Laplace, there can be only one which is bounded and non-negative for $0 \leqslant x < +\infty$. Since the function $\varphi_r(x)$ has both these properties, for its coincidence with function (33.1) it is sufficient to establish the positiveness of the latter. For this in turn, it is sufficient to show that $C_{rk} \geqslant 0$ $(0 \leqslant k \leqslant r)$. In his investigations, Palm gives explicit expressions for the numbers $a_{rk}$, and by an analysis of these expressions shows the positiveness of all coefficients $C_{rk}$. Thus, for all $r \geqslant 0$,

$$\varphi_r(x) = \sum_{k=0}^{r} C_{rk} \exp - a_{rk}x \qquad (0 \leqslant x < +\infty)$$

and Palm's problem may be regarded as completely solved. It may be seen that, for any $r$, Palm's function $\varphi_r(x)$, uniquely defining the stream of calls lost to $L_r$ (i.e. entering into $L_{r+1}$), is a linear combination of $r+1$ exponential functions. In particular, Palm presents the following expression for $\varphi_1(x)$ —

$$\varphi_1(x) = \left(\frac{1}{2} + \frac{1}{4\sqrt{\lambda + \frac{1}{4}}}\right) \exp\left[-\left(\lambda + \frac{1}{2} - \sqrt{\lambda + \frac{1}{4}}\right)x\right] +$$

$$\left(\frac{1}{2} - \frac{1}{4\sqrt{\lambda + \frac{1}{4}}}\right) \exp\left[-\left(\lambda + \frac{1}{2} + \sqrt{\lambda + \frac{1}{4}}\right)x\right].$$

# *PART THREE*

# SYSTEMS ALLOWING DELAY

## CHAPTER 9

## THE CASE OF AN EXPONENTIAL DISTRIBUTION FOR THE LENGTH OF CONVERSATIONS

### 34 Probabilities of different states

A fully-accessible collection allowing waiting will now be studied. An incoming call must wait for a conversation when and only when it finds all $n$ lines of a collection occupied. Thus here our former "probability of loss" (i.e. the probability of finding all the lines occupied) can be called the "probability of waiting". This variable understandably plays a part in the evaluation of the quality of the work of the collection. However, for a system with waiting this part is comparatively small, since, even if a significant majority of calls have to wait, the service must be acknowledged as completely satisfactory if the periods of delay are mostly very small. It is not the frequency of delay ("losses") which plays a decisive role but the nature of the time of waiting $\gamma$ as a random variable. The frequency of the delays ("losses") gives only one feature of this picture, the probability of the inequality $\gamma > 0$. Thus it can be seen that the final objective in the investigation of systems with delay is always *the distribution function of the time of waiting* $\gamma$.

For all problems that will be investigated it is immaterial whether the given collection of lines is ordered or not. It will always be assumed that an entering stream is simple with parameter $\lambda$. The calls are serviced in order of their arrival. The lengths of conversations will always be assumed to be independent of each other as also of the course of the stream of calls. As far as the distribution function of these lengths is concerned, this gives rise to the fundamental difference from problems of the theory of systems without delay. It is usually true that for different distributions of the lengths of conversations different methods need to be employed for investigating the time of waiting. The present chapter will be concerned with the simple instance where the lengths $l$ of the conversations are subject to an exponential law

$$P\{l > t\} = e^{-\beta t}, \qquad t > 0, \beta > 0 \text{ and is a constant.}$$

In this instance, a full solution of the problem has been given by Erlang [7].

$P_k(t)$ will be used to denote the probability that at the moment $t$ a system is in "state $k$", i.e. there are $k$ "present" (speaking or waiting) calls. When $k \leqslant n$, $k$ lines are occupied, and none are waiting. When $k > n$ all $n$ lines are occupied and there are $k - n$ waiting ones.

If $k < n$, similar circumstances exist as for systems with losses, since for $k < n$ there are neither losses nor waiting. All the results deduced in § 20 thus remain true and, as there, lead us to a system of equations

$$P'_0(t) = -\lambda P_0(t) + \beta P_1(t)$$
$$P'_k(t) = \lambda P_{k-1}(t) - (\lambda + k\beta)P_k(t) + (k+1)\beta P_{k+1}(t) \qquad (0 < k < n)$$

(cf. ($\mathscr{E}$), § 20, where we assumed $\beta = 1$). But the last equation of the system ($\mathscr{E}$) should now be replaced by another, since transference is now possible into state $n$ from state $n + 1$, which had no meaning for a system with losses.

Let us investigate what happens for any $k \geqslant n$. As before, $P_{rs}(\tau)$ will denote the probability of a transition of the system from state $r$ to state $s$ during time $\tau$, and the symbol $\approx$ will be used for an equation accurate to the infinitesimal order $o(\tau)$ as $\tau \to 0$. Then, by analogy with § 20, it follows that, for $k \geqslant n$ and as $\tau \to 0$

$$P_{k-1,k}(\tau) \approx \lambda\tau, \quad P_{k+1,k}(\tau) \approx n\beta\tau;$$
$$P_{kk}(\tau) \approx 1 - \lambda\tau - n\beta\tau, \quad P_{ik}(\tau) \approx 0 \ (|i-k| > 1)$$

(the difference from the situation when $k < n$ now arises since $P_{k+1,k}(\tau) \approx n\beta\tau$ and not $\approx (k+1)\beta\tau$ as before, because for $k \geqslant n$ in state $k + 1$ there are $n$ and not $k + 1$ lines occupied). These values, used with arguments exactly analogous to those used in § 20, lead to the relation

$$P_k(t+\tau) \approx P_{k-1}(t)\lambda\tau + P_k(t)\,(1-\lambda\tau-n\beta\tau) + P_{k+1}(t)n\beta\tau$$

from which, again, by close analogy with the argument of § 20, and with the aid of a passage to the limit, we get

$$P'_k(t) = \lambda P_{k-1}(t) - (\lambda + n\beta)P_k(t) + n\beta P_{k+1}(t) \qquad (k \geqslant n).$$

Thus the whole of system ($\mathscr{E}$) § 20 is now replaced by the (infinite) system of equations

$$\left.\begin{aligned} P'_0(t) &= -\lambda P_0(t) + \beta P_1(t) \\ P'_k(t) &= \lambda P_{k-1}(t) - (\lambda + k\beta)P_k(t) + (k+1)\beta P_{k+1}(t) \\ &\qquad\qquad\qquad\qquad\qquad\qquad (0 < k < n) \\ P'_k(t) &= \lambda P_{k-1}(t) - (\lambda + n\beta)P_k(t) + n\beta P_{k+1}(t) \qquad (k \geqslant n). \end{aligned}\right\} \quad (34.1)$$

As for a system with losses, probabilities of states are taken as limits to which the probabilities $P_k(t)$ tend as $t \to \infty$. For systems with losses the existence of these limits has been proved (Markoff's theorem, § 19). For systems allowing waiting (i.e. for equations 34.1) such a proof can also be derived. However, it is here much more complicated and requires a completely different set of ideas since Markoff's method is closely linked with the assumption of a finite number of possible states of the system. This proof will not be given here, only a reference to its possibility(*). One must observe further that for this new instance, in spite of the existence of limits of the variables $P_k(t)$ as $t \to \infty$, it is still necessary to prove the possibility of the passage to the limit in the whole of system (34.1). This question did not arise in § 20, as a finite system of equations was used there.

Thus it is assumed that as $t \to \infty$, the following limits exist —

$$\lim_{t\to\infty} P_k(t) = p_k \qquad (k = 0,1, \ldots)$$

and that a corresponding passage to the limit is simultaneously possible in all equations of the system (34.1). As the left-hand sides of all these equations in the limit are then changed to zero (this is precisely as proved in § 20), a system of linear equations is derived —

$$\left.\begin{aligned} -\lambda p_0 + \beta p_1 &= 0 \\ \lambda p_{k-1} - (\lambda + k\beta) p_k + (k+1)\beta p_{k+1} &= 0 \qquad (0 < k < n) \\ \lambda p_{k-1} - (\lambda + n\beta) p_k + n\beta p_{k+1} &= 0 \qquad (k \geqslant n). \end{aligned}\right\} \quad (34.2)$$

which, in conjunction with the normalising condition

$$\sum_{k=0}^{\infty} p_k = 1$$

serves to determine the quantities $p_k$.

Assuming that

$$\lambda p_{k-1} - k\beta p_k = z_k \qquad (1 \leqslant k \leqslant n),$$

system (34.2) gives

$$z_1 = 0, \quad z_k - z_{k+1} = 0 \qquad (1 \leqslant k < n)$$

from which

$$z_k = 0 \qquad (0 < k \leqslant n).$$

If, for brevity, it is assumed that $\lambda/\beta = y$, then

$$p_k = \frac{y^k}{k!}\, p_0 \qquad (0 \leqslant k \leqslant n)$$

and, in particular for $k = n$,

$$p_n = \frac{y^n}{n!}\, p_0. \tag{34.3}$$

To find the value of $p_n$ for $k > n$ the last group of equations in (34.2) must be employed. Putting them in the form

$$n\beta(p_{k+1} - p_k) = \lambda(p_k - p_{k-1}) \qquad (k \geqslant n)$$

(*)I do not know of any published exposition of this proof, although the problem itself has been examined many times in this way (Erlang [7], Fry [4], Kolmogoroff [2], Feller [3]). Erlang brings in the possibility of the passage to the limit as a special postulate. The other authors confine themselves to a short demonstration of the possibility of a proof.

and summing over $k$ from $n$ to $n + r$,

$$n\beta(p_{n+r+1} - p_n) = \lambda(p_{n+r} - p_{n-1}),$$

from which

$$n\beta\, p_{n+r+1} + z_n = \lambda\, p_{n+r},$$

or, since $z_n = 0$, $\lambda/\beta = y$,

$$p_{n+r+1} = \frac{y}{n}\, p_{n+r} \qquad (r \geqslant 0).$$

Thus, by virtue of (34.3)

$$p_{n+r+1} = \left(\frac{y}{n}\right)^{r+1} \frac{y^n}{n!}\, p_0 \qquad (r \geqslant 0)$$

and for any $k \geqslant n$,

$$p_k = \left(\frac{y}{n}\right)^{k-n} p_n = \frac{y^k}{n!n^{k-n}}\, p_0. \tag{34.4}$$

Combining this result with that obtained previously for $k \leqslant n$, we find

$$p_k = \frac{y^k}{k!}\, p_0 \qquad (0 \leqslant k \leqslant n),$$

$$p_k = \frac{y^k}{n!n^{k-n}}\, p_0 \qquad (k \geqslant n). \tag{34.5}$$

It is still necessary to find $p_0$. The normalising condition

$$\sum_{k=0}^{\infty} p_k = 1$$

gives

$$\frac{1}{p_0} = \sum_{k=0}^{n-1} \frac{y^k}{k!} + \frac{n^n}{n!} \sum_{k=n}^{\infty} \left(\frac{y}{n}\right)^k = s_{n-1}(y) + \frac{y^n}{(n-1)\,!\,(n-y)},$$

where $s_m(y) = \sum_{k=0}^{m} (y^k/k!)$. The analogue of this equation for a system with losses was (§ 20)

$$\frac{1}{p_0} = s_n(y) = s_{n-1}(y) + \frac{y^n}{n!}.$$

Further, the probability of finding all lines occupied (the "probability of waiting") is, in consequence of (34.4) equal to

$$\pi = \sum_{k=n}^{\infty} p_n = \frac{n^n p_0}{n!} \sum_{k=n}^{\infty} \left(\frac{y}{n}\right)^k = \frac{y^n}{n!} \frac{p_0}{1 - \dfrac{y}{n}}. \tag{34.6}$$

## 35 The distribution function of the waiting time

Now the probability $P\{\gamma > t\}$ may be found that any call entering at a randomly chosen moment has a waiting time greater than $t$. Denote by $P_k\{\gamma > t\}$ the conditional probability of this inequality on the assumption that the said call has found the system in state $k$. By the formula of compound probability

$$P\{\gamma > t\} = \sum_{k=0}^{\infty} p_k P_k\{\gamma > t\}$$

or, since $P_k\{\gamma > t\} = 0$ for $k < n$ and $t \geqslant 0$,

$$P\{\gamma > t\} = \sum_{k=n}^{\infty} p_k P_k\{\gamma > t\}. \tag{35.1}$$

The variables $p_k$ are known. It remains to determine the variables $P_k\{\gamma > t\}$ for all $k \geqslant n$.

Assume that $k - n = \nu$ ($\nu = 0,1,2,\ldots$). The problem is to find the probability of the inequality $\gamma > t$ when all the lines are occupied at the moment of the call and, in addition, there are $\nu$ calls waiting. Under these circumstances our call starts a conversation after the $(\nu + 1)$th freeing of a line. The required probability is therefore the probability that during time $t$ after the occurrence of our call there will occur not more than $\nu$ freeings of a line. Suppose $q_r(t)$ $(0 \leqslant r \leqslant \nu)$ is the probability that during this time there occur precisely $r$ freeings. Then, since $k - n = \nu$

$$P_k\{\gamma > t\} = \sum_{r=0}^{k-n} q_r(t) \qquad (k \geqslant n).$$

But the stream of freeings during the time that our call is waiting represents, by virtue of the exponential law of the lengths of conversations, a simple stream with parameter $n\beta$ (since the probability that no freeing will occur during time $t$ from the moment when all lines are occupied is equal to $(e^{-\beta t})^n = e^{-\beta nt}$). The magnitude $q_r(t)$ is the probability that during time $t$ there will occur $r$ events of this stream. Thus, using the formulae from Chapter 1,

$$q_r(t) = e^{-n\beta t}\frac{(n\beta t)^r}{r!} \qquad (0 \leqslant r \leqslant \nu),$$

and

$$P_k\{\gamma > t\} = \sum_{r=0}^{k-n} e^{-n\beta t}\frac{(n\beta t)^r}{r!} \qquad (k \geqslant n).$$

Returning to formula (35.1) and using equation (34.4) now gives

$$\begin{aligned}
P\{\gamma > t\} &= \sum_{k=n}^{\infty} p_k \sum_{r=0}^{k-n} e^{-n\beta t}\frac{(n\beta t)^r}{r!} \\
&= e^{-n\beta t}\sum_{k=n}^{\infty}\left(\frac{y}{n}\right)^{k-n} p_n \sum_{r=0}^{k-n}\frac{(n\beta t)^r}{r!} \\
&= p_n e^{-n\beta t}\sum_{r=0}^{\infty}\frac{(n\beta t)^r}{r!}\sum_{k=n+r}^{\infty}\left(\frac{y}{n}\right)^{k-n} \\
&= p_n e^{-n\beta t}\sum_{r=0}^{\infty}\frac{(n\beta t y)^r}{n^r r!}\sum_{k=n+r}^{\infty}\left(\frac{y}{n}\right)^{k-n-r}
\end{aligned}$$

$$= \frac{p_n e^{-n\beta t}}{1 - \frac{y}{n}} \sum_{r=0}^{\infty} \frac{(\lambda t)^r}{r!} = \frac{p_n}{1 - \frac{y}{n}} e^{-(n\beta - \lambda)t}$$

or, since from (34.3) and (34.6) $p_n = \frac{y^n}{n!} p_0 = \pi \left(1 - \frac{y}{n}\right)$,

$$P\{\gamma > t) = \pi e^{-(n\beta - \lambda)t} \qquad (t \geqslant 0).$$

This solves the problem. It may be seen that under the stated conditions the waiting time follows an exponential law of distribution with parameter $n\beta - \lambda$. In addition,

$$P\{\gamma > 0\} = \pi$$

as it should be ($\pi$ was used at the end of § 34 to denote the probability of finding all lines occupied, i.e. the probability of waiting).

CHAPTER 10

# SINGLE-LINE SYSTEMS WITH A FIXED LENGTH OF CONVERSATION

## 36 Difference-differential equations for the problem

Without the restriction of an exponential distribution for the lengths of conversations, the investigation of systems allowing waiting is beset with great difficulties. One can succeed in getting simple and conclusive results only in certain particular situations. From a practical point of view, the case of systems with a single line is of special importance. For these, the problem of investigation of the time of waiting can be further simplified by statistical propositions of a very general nature. This particular instance will be considered in the remainder of the book.

In the present chapter, the theory of single-line systems will be explained on the assumption that all conversations have exactly the same length $\tau$. This theory was created by Erlang and is as interesting for the originality of its methods as for the completeness of its results. In all other respects, the statistical assumptions of preceding chapters will be retained.

Suppose that any two successive calls are observed. Let $\gamma_0$ be the time of waiting of the first and $\gamma$ that of the second. We shall denote by $z$ the distance between these two calls. It is clear that if $z \geqslant \gamma_0 + \tau$, then at the time of the second call the (only) line is free and $\gamma = 0$. Conversely, if $z < \gamma_0 + \tau$, then when the second call occurs the line is occupied and it must wait for a period of time $\gamma = \gamma_0 + \tau - z$. Thus, for a fixed $z = u$,

$$\gamma = \begin{cases} 0 & (\gamma_0+\tau-u \leqslant 0) \\ \gamma_0 + \tau - u & (\gamma_0+\tau-u \geqslant 0) \end{cases}$$

or, what is the same thing,

$$\gamma = \max(0, \gamma_0+\tau-u).$$

Suppose that $t$ is any positive number. It will now be established that for a fixed $z = u$, the inequalities $\gamma < t$ and $\gamma_0 < t + u - \tau$ are equivalent. In fact, if $\gamma < t$ then either $\gamma = 0$ and hence $\gamma_0 \leqslant u - \tau < u + t - \tau$, or $\gamma = \gamma_0 + \tau - u$ and hence $\gamma_0 + \tau - u < t$ and $\gamma_0 < t + u - \tau$. Conversely, if $\gamma_0 < t + u - \tau$ then either $\gamma_0 \leqslant u - \tau$, $\gamma_0 + \tau - u \leqslant 0$, $\gamma = 0 < t$, or $\gamma_0 > u - \tau$, $\gamma = \gamma_0 + \tau - u < t$. Thus, if $P_u$ denotes the conditional probability calculated on the assumption that $z = u$, then for any $u > 0$,

$$P_u\{\gamma < t\} = P_u\{\gamma_0 < t + u - \tau\} \qquad (t > 0).$$

But $\gamma_0$ (the waiting time of the first call) does not depend (as a random variable) on when the second call follows, i.e. the quantity $z$ can have any value. Thus the conditional probability $P_u\{\gamma_0 < t + u - \tau\}$ of the inequality $\gamma_0 < t + u - \tau$ is equal to the unconditional probability $P\{\gamma_0 < t + u - \tau\}$ of the same inequality, and

$$P_u\{\gamma < t\} = P\{\gamma_0 < t + u - \tau\} \qquad (t > 0). \tag{36.1}$$

Denote by $f(t)$ the distribution function of the variable $\gamma$, assuming

$$f(t) = P\{\gamma < t\}.$$

Since the stream of calls is assumed to be simple with parameter $\lambda$, the probability of the inequalities $u < z < u + du$ (accurate to higher orders of the infinitesimal) is equal to $\lambda e^{-\lambda u} du$ and, from the formula of compound probability,

$$f(t) = \int_0^\infty \lambda e^{-\lambda u} P_u\{\gamma < t\} du$$

or by virtue of (36.1) for $t > 0$

$$f(t) = \int_0^\infty \lambda e^{-\lambda u} P\{\gamma_0 < t + u - \tau\} du \qquad (t > 0).$$

But the distribution function of the variable $\gamma_0$ is given by the same function $f(t)$ as that of the variable $\gamma$. Thus,

$$f(t) = \int_0^\infty \lambda e^{-\lambda u} f(t+u-\tau) du \qquad (t > 0).$$

This equation serves as a basis for determining the required function $f(t)$. First of all, assuming this function to be differentiable, it follows that

$$f'(t) = \int_0^\infty \lambda e^{-\lambda u} f'(t+u-\tau) du \qquad (t > 0).$$

Integration by parts gives

$$f'(t) = \Big[\lambda e^{-\lambda u} f(t+u-\tau)\Big]_0^\infty + \lambda^2 \int_0^\infty e^{-\lambda u} f(t+u-\tau) du$$
$$= -\lambda f(t-\tau) + \lambda f(t),$$

or

$$f'(t) = \lambda[f(t) - f(t-\tau)] \qquad (t > 0).$$

Thus for determining the function $f(t)$ a difference-differential equation of a simple type is derived. This equation can be still further simplified by a transformation of the required function —

$$f(t) = e^{\lambda t} g(t)$$

which gives for the new unknown function the equation

$$g'(t) = -\lambda e^{-\lambda \tau} g(t-\tau) \qquad (t > 0). \tag{36.2}$$

## 37 Distribution function of the waiting time

First, consider a segment of time $0 < t \leqslant \tau$. Since $t > 0$, all the relations deduced in § 36 hold. But, for $t \leqslant \tau, f(t-\tau) = P\{\gamma < t - \tau\} = 0$ and consequently the relation (36.2) gives

$$g'(t) = 0, \quad g(t) = c = \text{const.} \qquad (0 < t \leqslant \tau),$$

from which

$$f(t) = ce^{\lambda t} \qquad (0 < t \leqslant \tau).$$

To find the constant $c$, let $t$ tend to zero in the last equation. In the limit, $c = f(+0)$, but $f(-0) = 0$ and so $c = f(+0) - f(-0) = P\{\gamma = 0\}$. This is the probability that a call occurring at any arbitrarily chosen moment will not have to wait (the line is free). Thus $c = 1 - \alpha$, where $\alpha$ denotes the probability of finding a line occupied at an arbitrarily chosen moment. In other words, $\alpha$ is the mathematical expectation of the total length of all conversations occurring in the course of a unit of time. But the mathematical expectation of the number of conversations in a unit of time is equal to $\lambda$ and the length of each conversation is equal to $\tau$. Consequently

$$\alpha = \lambda\tau, \quad c = 1 - \alpha = 1 - \lambda\tau.$$

Thus, finally

$$f(t) = (1-\alpha)e^{\lambda t} \qquad (\alpha = \lambda\tau, \qquad 0 < t \leqslant \tau)$$

and the distribution function of the waiting time has been found for the segment $0 < t \leqslant \tau$.

It will now be shown that *for any non-negative integer $n$ the function $g(t)$ is determined in the segment $n\tau < t \leqslant (n+1)\tau$ by*

$$g(t) = (1-\alpha) \sum_{k=0}^{n} e^{-\lambda k\tau} \frac{(k\tau-t)^k}{k!} \lambda^k. \tag{37.1}$$

Since the function $f(t)$ is easily found from function $g(t)$, a clear expression for the distribution function of the waiting time is provided.

For $n = 0$, formula (37.1) gives

$$g(t) = 1 - \alpha \qquad (0 < t \leqslant \tau),$$

as proved above. Thus formula (37.1) can be proved by induction. Assume that it is true for any integer $n \geqslant 0$ and, in such an instance, it can be shown that it remains true also for the integer $n + 1$.

Thus, assuming the relation (37.1) to be true for $n\tau < t \leqslant (n+1)\tau$ if $(n+1)\tau < t \leqslant (n+2)\tau$ then, by virtue of (36.2)

$$g'(t) = -\lambda e^{-\lambda\tau} g(t-\tau)$$

$$= -\lambda e^{-\lambda\tau} (1-\alpha) \sum_{k=0}^{n} e^{-\lambda k\tau} \frac{(k\tau-t+\tau)^k}{k!} \lambda^k.$$

Integrating this equation over $t$ from $(n+1)\tau$ to a value which we shall again denote by $t$, it is found that

$$g(t) - g[(n+1)\tau] = -\lambda e^{-\lambda\tau}(1-\alpha)\sum_{k=0}^{n}\frac{e^{-\lambda k\tau}\lambda^k}{k!}\int_{(n+1)\tau}^{t}[(k+1)\tau - u]^k du.$$

Here,

$$\int_{(n+1)\tau}^{t}[(k+1)\tau - u]^k du$$

$$= -\frac{1}{k+1}\{[(k+1)\tau - t]^{k+1} - [(k-n)\tau]^{k+1}\},$$

and so

$$g(t) - g[(n+1)\tau] = \lambda e^{-\lambda\tau}(1-\alpha)\sum_{k=0}^{n}\frac{e^{-\lambda k\tau}\lambda^k}{(k+1)!}\{[(k+1)\tau - t]^{k+1} - [(k+1)\tau - (n+1)\tau]^{k+1}\}$$

$$= (1-\alpha)\sum_{k=0}^{n}\frac{e^{-\lambda(k+1)\tau}\lambda^{k+1}}{(k+1)!}\{[(k+1)\tau - t]^{k+1} - [(k+1)\tau - (n+1)\tau]^{k+1}\}$$

$$= (1-\alpha)\sum_{r=1}^{n+1}\frac{e^{-\lambda r\tau}\lambda^r}{r!}\{[r\tau - t]^r - [r\tau - (n+1)\tau]^r\}$$

$$= (1-\alpha)\sum_{r=0}^{n+1}\frac{\lambda^r e^{-\lambda r\tau}}{r!}(r\tau - t)^r$$

$$- (1-\alpha)\sum_{r=0}^{n+1}\frac{\lambda^r e^{-\lambda r\tau}}{r!}[r\tau - (n+1)\tau]^r, \qquad (37.2)$$

where it is possible to start the summation for $r$ from zero, since the terms for $r = 0$ in both sums cancel one another.

Formula (37.1), according to our assumption, is true for $n\tau < t \leqslant (n+1)\tau$. Setting $t = (n+1)\tau$ in it, we find

$$g[(n+1)\tau] = (1-\alpha)\sum_{r=0}^{n+1}\frac{\lambda^r e^{-\lambda r\tau}}{r!}[r\tau - (n+1)\tau]^r \qquad (37.3)$$

where the summation extends up to $n + 1$, because the term for $r = n + 1$ is equal to zero. Finally, combining equations (37.2) and (37.3) it is found that

$$g(t) = (1-\alpha)\sum_{r=0}^{n+1}\frac{\lambda^r e^{-\lambda r\tau}}{r!}(r\tau - t)^r$$

for any $t$ in the period $(n+1)\tau < t \leqslant (n+2)\tau$, which was to be proved

Thus, for any $t > 0$,

$$P\{\gamma < t\} = f(t) = e^{\lambda t}(1-\lambda\tau)\sum_{k=0}^{n} e^{-\lambda k\tau}\frac{(k\tau - t)^k}{k!}\lambda^k$$

where the non-negative integer $n$ is defined by the inequalities

$$n\tau < t \leqslant (n+1)\tau.$$

# CHAPTER 11

# GENERAL THEORY OF SINGLE-LINE SYSTEMS

## 38 Statement of the problem and definitions

In the preceding two chapters, following Erlang's classic work, the distribution function of the waiting time has been found under two of the simplest assumptions concerning the distribution function of the length of conversation — the instance of an exponential distribution (*vide* Chapter 9) and (for one-line systems) the instance of a fixed length of conversation (*vide* Chapter 10). However, in practice, these two simple distributions occur only on rare occasions. In the majority of applications, these may provide at best only an approximation to the real distribution of the length of conversations. Further, even for such a calculation, no basis exists in many instances. It is desirable to devise a method of determining the distribution function of the waiting time (if only the most important of its statistical characteristics) under broad assumptions regarding the distribution of the length of conversations.

A general statement of this problem leads to calculations which are difficult to relate on account of their complexity. Thus in what follows our attention will be concentrated upon the important practical case of a *system with a single line.* As regards the distribution function of the length of conversations we shall confine ourselves to the natural requirements of the existence of a finite mathematical expectation, leaving this law hereafter completely arbitrary. It will be seen that, in these circumstances, the problem of finding a distribution function for the time of waiting is decided in principle by comparatively simple methods.

The rest of the present chapter will be concerned with a collection of one line into which a simple stream of calls with parameter $\lambda$, enters. Calls finding the line occupied wait for it to become free and occupy it in the order of their reception. The lengths of conversations depend neither on one another nor on the number of waiting calls. The probability that the length of any arbitrarily chosen conversation will be greater than $t$ will be denoted by $F(t)$, so that the mean length of a conversation will be

$$-\int_0^\infty t dF(t) = \int_0^\infty F(t)dt = s.$$

In addition, the following notation will be used throughout the chapter —

$a$ is the probability that at any arbitrarily chosen moment of time the line will be occupied; in other words the mathematical expectation of the total time of occupation of the line during one hour (an hour will be conditionally adopted as the unit of time);

$\pi_k$ $(k = 0,1,2,\ldots)$ is the probability that at the start of any conversation there are $k$ others waiting;

$v_k(t) = e^{-\lambda t}(\lambda t)^k/k!$ is the probability that during the course of a period of time of length $t$ there will occur $k$ calls;

$\gamma$ (random variable) is the waiting time for a call which has occurred at an arbitrarily chosen moment of time.

Further symbols will be explained as they are introduced.

## 39 Subsidiary hypotheses

*Lemma* 1. *Suppose that* $0 < a < b$. *The probability of finding the line occupied at any moment of time by a conversation of length between a and b is equal to*

$$-\lambda\int_a^b u dF(u).$$

*Proof.* Denote by $P_T(a,b)$ the probability of finding the line occupied by a conversation of length between $a$ and $b$(*), at a moment of time arbitrarily chosen within the period $(0,T)$. Lemma 1 thus states that

$$\lim_{T\to\infty} P_T(a,b) = -\lambda\int_a^b u dF(u).$$

$L_T^*(a,b)$ will be used hereafter to denote the total length of all conversations and parts of conversations of length $(a,b)$ occurring in the period $(0,T)$. $L_T^*(a,b)$ is a random variable. If it takes any value $l$, then the corresponding conditional probability of finding a conversation of length $(a,b)$ is equal to $l/T$. Thus, by virtue of the formula of compound probability,

$$P_T(a,b) = \sum_l P\{L_T^*(a,b) = l\}\frac{l}{T} = \frac{1}{T} ML_T^*(a,b),$$

where $M$ is the symbol of mathematical expectation.

For the proof of lemma 1 it is thus sufficient to establish that, as $T\to\infty$,

$$\lim \frac{1}{T} ML_T^*(a,b) = -\lambda\int_a^b u dF(u). \tag{39.1}$$

Suppose that $L_T(a,b)$ denotes the total length of conversations of length $(a,b)$ beginning in the segment $(0,T)$. Since

(*)Hereafter, for the sake of brevity, we shall call such a conversation a "conversation of length $(a,b)$".

$$|L_T(a,b) - L_T^*(a,b)| < b,$$

and consequently also

$$|ML_T(a,b) - ML_T^*(a,b)| < b,$$

it follows that (39.1) is equivalent to the relation

$$\frac{1}{T} ML_T(a,b) \to -\lambda \int_a^b u dF(u) \qquad (T \to \infty).$$

But, since $ML_T(a,b)$ is proportional to $T$, the last relation simply means that for any $T > 0$,

$$\frac{1}{T} ML_T(a,b) = -\lambda \int_a^b u dF(u).$$

In particular, this relation will be established (and thus lemma 1 proved) if it can be shown that[*]

$$\lim_{T \to \infty} \frac{1}{T} ML_T(a,b) = -\lambda \int_a^b u dF(u). \tag{39.2}$$

Suppose $u$ to be any positive number and $T < u$[†]. Then not more than one conversation of length $(u,u+du)$ can begin in the period $(0,T)$, so that the random variable $L_T(u,u+du)$ can, apart from the value 0, bear only values of the size $u + \theta du$, where $0 < \theta < 1$. Such a value will be borne by it if — (1) at least one conversation starts in the segment $(0,T)$, and (2) the last of the conversations starting in $(0,T)$ has the length $(u,u+du)$. The probability of (2) is $F(u) - F(u+du)$ and the probability of (1) as $T \to 0$ has the form $\lambda T + o(T)$, where $\lambda$ is the mean number of conversations starting in a unit of time, coinciding with the parameter of the incoming stream of calls. Thus the random variable $L_T(u,u+du)$ for the case under consideration is either equal to zero, or has a value of the size $u+\theta du$, in which the latter has a probability $[\lambda T + o(T)]\ [F(u) - F(u+du)]$. Thus

$$ML_T(u,u+du) = (u+\theta du)\,[\lambda T + o(T)]\,[F(u) - F(u+du)]$$

and consequently, summing for $u$ from $a$ to $b$,

$$ML_T(a,b) = [-\lambda T + o(T)] \int_a^b u dF(u).$$

(39.2) follows from this, and lemma 1 also.

(*) See Additional Note 3 on page 122.

(†) See Additional Note 4 on page 122.

Lemma 1 is therefore proved. For $a = 0$, $b = +\infty$ the probability under investigation is the probability $\alpha$ that at an arbitrarily chosen moment of time the line will be occupied. The important formula

$$\alpha = -\lambda \int_0^\infty t dF(t) = \lambda s$$

is thus derived (which it would also be possible to realise by a direct argument).

A conversation at the start of which there are $k$ others waiting will be called *a conversation of the type "k"* ($k = 0,1,2, \ldots$) so that the magnitude $\pi_k$ which was introduced in § 38 may be defined as the proportion of the conversations of type $k$ among all incoming conversations (the probability that an arbitrarily chosen conversation will be a conversation of type $k$).

*Lemma* 2. *If at a given moment the line is occupied, then the conditional probability that it will be occupied by a conversation of type k is equal to* $\pi_k$.

*Proof.* On average, amongst $\lambda$ conversations entering in a unit of time, $\lambda\pi_k$ will be conversations of type $k$. Since the length of the conversation as a random variable is independent of the number of those waiting at its start (i.e. of its type), the mean length of the conversation of type $k$ is equal to $s$, and consequently the total length of conversations of type $k$ occurring in the unit of time is on average equal to $\lambda\pi_k s = \alpha\pi_k$. But the conditional probability with which lemma 2 is concerned is the ratio of this mean total length to the time of occupation of the line, i.e. to $\alpha$. Thus this conditional probability is equal to $\pi_k$ and lemma 2 is proved.

Now, select any conversation of length $t > 0$ and find the probability $u_k(z)$ $(0 < z \leqslant t)$ that after a period of time $z$ after the start of this conversation the number of others waiting will be equal to $k$ ($k = 0,1,2, \ldots$). Since the length and type of the conversation are mutually independent, the probability that at the beginning of the chosen conversation there will be $r$ others waiting is equal to $\pi_r$. If this occurs, then for the number of waiting calls to reach $k$ during time $z$ it is necessary and sufficient that $r \leqslant k$, and that during this period of time $z$ there should occur $k - r$ new calls, the probability of which is

$$v_{k-r}(z) = e^{-\lambda z} \frac{(\lambda z)^{k-r}}{(k-r)!}.$$

The following results are thus arrived at —

*Lemma* 3. *The probability* $u_k(z)$ *that at the end of a period of time* $z \leqslant t$ *after the beginning of an arbitrarily chosen conversation of length t there are k others waiting is equal to*

$$u_k(z) = \sum_{r=0}^{k} \pi_r v_{k-r}(z) \qquad (k = 0,1,2, \ldots).$$

Now $\tau$ will be used to denote the period of time between an arbitrarily chosen moment and the termination of the conversation at that moment (the remaining length of a conversation chosen at random). It may be then stated —

*Lemma* 4. *For* $k \geqslant 0$, $x > 0$ *and small* $dx > 0$, *the probability of finding at a randomly chosen moment a conversation with* $x < \tau < x + dx$ *and k others waiting is equal to*

$$-\lambda dx \int_x^\infty u_k(t-x)dF(t) + o(dx).$$

*Proof.* A difficulty in establishing lemma 4 consists in the fact that the distribution function of the random variable $\tau$ depends upon how many others are waiting at the chosen moment of time. If we denote by

(*a*) the event consisting of the fact that there are $k$ others waiting at the chosen moment, and by

(*b*) the event that at the chosen moment there occurs a conversation with $x < \tau < x + dx$,

these two events are interdependent. Our problem is to determine their joint probability $P(a,b)$. With this objective, the conditional probability $P_t(a,b)$ of these two events, assuming that the chosen conversation has length $t$, will be initially investigated. We can write

$$P_t(a,b) = P_t(b)P_{tb}(a). \qquad (39.3)$$

Since the chosen moment, if it falls in a conversation of length $t$, can occur with equal probability at any moment of this conversation, then

$$P_t(b) = \left\{ \begin{matrix} dx/t & (t > x) \\ 0 & (t \leqslant x) \end{matrix} \right\}. \qquad (39.4)$$

On the other hand, event $b$ is equivalent to the event

$$t - x - dx < t - \tau < t - x$$

where $t - \tau$ is the elapsed part of the chosen conversation. Thus $P_{tb}(a)$ is the conditional probability that, at any arbitrarily chosen moment, $k$ others will be waiting if it is known that the conversation at that moment is of length $t$ and with an elapsed length of between $t - x - dx$ and $t - x$. From the definition of the function $u_k(x)$ (*vide* lemma 3) this probability, as $dx \to 0$, is asymptotically equal to $u_k(t-x)$(*).

(*)The continuity of $u_k(z)$ follows from lemma 3.

Thus, by virtue of (39.3) and (39.4), we have

$$P_t(a,b) = \begin{cases} u_k(t-x)dx/t + o(dx) & (t > x) \\ 0 & (t \leqslant x). \end{cases}$$

This is the conditional probability of the occurrence of events $a$ and $b$ at the same place, *assuming that the chosen conversation is of length* $t$ — it remains now to eliminate this condition. From lemma 1, the distribution function of the length of the chosen conversation is

$$G(t) = -\lambda \int_t^{\infty} udF(u),$$

and by the formula of compound probability we have

$$\begin{aligned} P(a,b) &= -\int_0^{\infty} P_t(a,b)dG(t) \\ &= -\int_x^{\infty} u_k(t-x)\frac{dx}{t}\,dG(t) + o(dx) \\ &= -\lambda \int_x^{\infty} u_k(t-x)\frac{dx}{t}\,tdF(t) \\ &= -\lambda dx \int_x^{\infty} u_k(t-x)dF(t) + o(dx), \end{aligned}$$

which was to be proved.

*Lemma* 5.

$$-\int_0^{\infty} u_{k+1}(x)dF(x) = \pi_k \qquad (k > 0)$$

$$-\int_0^{\infty} u_1(x)dF(x) = \pi_0\left\{1 + \int_0^{\infty} e^{-\lambda x}\,dF(x)\right\}.$$

*Proof.* From the meaning of the function $u_{k+1}(x)$ (*vide* the formulation of lemma 3) the integral $-\int_0^{\infty} u_{k+1}(x)dF(x)$ denotes the probability that at the end of the arbitrarily chosen conversation there are $k+1$ others waiting. In the case $k > 0$ this is equivalent to saying that the following conversation will begin with $k$ others waiting, and this probability is equal to $\pi_k$. With this the first of the above two equations is proved.

In the case $k = 0$ the probability that the chosen conversation will conclude with one other waiting will be *less* than $\pi_0$. The fact is that a conversation without others waiting can begin not only at the end of a previous conversation with one other waiting, but also if a call occurs at the moment when the line is free. Thus

$$-\int_0^\infty u_1(x)dF(x) = \pi_0 - \varrho,$$

where $\varrho > 0$. In order to determine $\varrho$, all the established relations need to be combined, and using the equations

$$\sum_{k=0}^\infty \pi_k = 1 \quad \text{and} \quad \sum_{k=0}^\infty u_k(x) = 1$$

for any $x$, gives

$$-\int_0^\infty [1 - u_0(x)]\, dF(x) = 1 - \varrho.$$

Since, by virtue of lemma 3, $u_0(x) = \pi_0 v_0(x) = \pi_0 e^{-\lambda x}$, it follows that

$$\varrho = -\pi_0 \int_0^\infty e^{-\lambda x}\, dF(x) \tag{39.5}$$

and lemma 5 is fully proved.

*Lemma* 6. *For* $|a| \leqslant 1$, $z > 0$

$$\sum_{k=0}^\infty a^k u_k(z) = e^{\lambda z(a-1)} \sum_{k=0}^\infty \pi_k a^k.$$

*Proof*. By virtue of lemma 3,

$$\sum_{k=0}^\infty a^k u_k(z) = \sum_{k=0}^\infty a^k \sum_{r=0}^k \pi_r e^{-\lambda z} \frac{(\lambda z)^{k-r}}{(k-r)!} = e^{-\lambda z} \sum_{r=0}^\infty \pi_r \sum_{k=r}^\infty a^k \frac{(\lambda z)^{k-r}}{(k-r)!}$$

$$= e^{-\lambda z} \sum_{r=0}^\infty \pi_r a^r \sum_{k=r}^\infty \frac{(a\lambda z)^{k-r}}{(k-r)!} = \exp\,[\lambda z(a-1)] \sum_{k=0}^\infty \pi_k a^k$$

which was to be proved.

Denote by $\psi(\xi)$, the characteristic function of the length of a conversation, i.e. suppose that

$$\psi(\xi) = -\int_0^\infty e^{i\xi t}\, dF(t).$$

Suppose further that, for any real $\xi$,

$$\chi(\xi) = \sum_{k=0}^\infty \pi_k [\psi(\xi)]^k.$$

Finally, let

$$\eta = \eta(\xi) = \frac{\lambda}{i}\,[\psi(\xi) - 1].$$

*Lemma* 7.

$$\chi(\xi) = (1-\alpha)\ \frac{1 - \psi(\xi)}{\psi(\eta) - \psi(\xi)}.$$

*Proof*. By virtue of lemma 5 we have (defining $\varrho$ by the formula (39.5))

$$\chi(\xi) = \sum_{k=0}^{\infty} [\psi(\xi)]^k \pi_k = -\sum_{k=0}^{\infty} [\psi(\xi)]^k \int_0^\infty u_{k+1}(x) dF(x) + \varrho$$

$$= \varrho - \frac{1}{\psi(\xi)} \int_0^\infty \left\{ \sum_{k=1}^{\infty} [\psi(\xi)]^k u_k(x) \right\} dF(x)$$

from which by virtue of lemma 6

$$\chi(\xi) = \varrho - \frac{1}{\psi(\xi)} \int_0^\infty \left\{ e^{\lambda x[\psi(\xi)-1]} \sum_{k=0}^{\infty} \pi_k [\psi(\xi)]^k - u_0(x) \right\} dF(x)$$

$$= \varrho - \frac{1}{\psi(\xi)} \int_0^\infty \left\{ e^{\lambda x[\psi(\xi)-1]}\, \chi(\xi) - \pi_0 e^{-\lambda x} \right\} dF(x)$$

$$= \varrho - \frac{\chi(\xi)}{\psi(\xi)} \int_0^\infty e^{\lambda x[\psi(\xi)-1]} dF(x) + \frac{1}{\psi(\xi)} \int_0^\infty \pi_0 e^{-\lambda x} dF(x)$$

$$= \varrho - \frac{\chi(\xi)}{\psi(\xi)} \int_0^\infty e^{i\eta x} dF(x) - \frac{\varrho}{\psi(\xi)}$$

$$= \varrho\, \frac{\psi(\xi) - 1}{\psi(\xi)} + \frac{\chi(\xi)\psi(\eta)}{\psi(\xi)}.$$ Thus

$$\chi(\xi)\, \frac{\psi(\xi) - \psi(\eta)}{\psi(\xi)} = \varrho\, \frac{\psi(\xi) - 1}{\psi(\xi)}$$ and

$$\chi(\xi) = \varrho\, \frac{1 - \psi(\xi)}{\psi(\eta) - \psi(\xi)}.$$

To prove lemma 7, it remains to show that $\varrho = 1 - \alpha$. But $\varrho$ is independent of $\xi$ and since $\chi(0) = 1$, the last formula gives

$$\varrho = \lim_{\xi \to 0} \frac{\psi(\eta) - \psi(\xi)}{1 - \psi(\xi)},$$

or, by L'Hôpital's rule

$$\varrho = \lim_{\xi \to 0} \frac{\psi'(\eta)\,\dfrac{d\eta}{d\xi} - \psi'(\xi)}{-\,\psi'(\xi)} = 1 - \lim_{\xi \to 0} \frac{\psi'(\eta)}{\psi'(\xi)}\, \frac{d\eta}{d\xi}\,.$$

But $\psi'(0) = is$, $\frac{d\eta}{d\xi} = \frac{\lambda}{i}\psi'(\xi) \to \lambda s = \alpha$ as $\xi \to 0$ and hence $\varrho = 1 - \alpha$.

## 40 The characteristic function of the waiting time

The object of our investigation is to find the distribution function of the random variable $\gamma$, the waiting time of a call occurring at an arbitrarily chosen moment of time. Since each distribution function is uniquely determined by a corresponding characteristic function, and since it is usually easier to find the most important properties of a random variable from its characteristic function than from its distribution function, our objective will be achieved by finding the characteristic function $\varphi(\xi)$ of the variable $\gamma$, i.e. the mathematical expectation of the expression $e^{i\gamma\xi}$ as a function of the real parameter $\xi$. This is what we shall try to do.

Hereafter let $M$ be the mathematical symbol for expectation, so that $\varphi(\xi) = Me^{i\gamma\xi}$. The following possibilities may be distinguished. First of all, at an arbitrarily chosen moment the line may be free. The probability of this is $1 - \alpha$, and in this case certainly $\gamma = 0$, $e^{i\gamma\xi} = 1$. Secondly, the line may be occupied with $k$ calls waiting ($k = 0,1,2, \ldots$). The probability of this will be denoted by $P(k)$ and the mathematical expectations calculated in this circumstance will be denoted by $M_k$. We then have

$$\varphi(\xi) = Me^{i\gamma\xi} = 1 - \alpha + \sum_{k=0}^{\infty} P(k) M_k\, e^{i\gamma\xi}.$$

If our call finds the line occupied with $k$ calls waiting, the time of waiting $\gamma$ consists of two parts — (1) the remaining length of that conversation which is proceeding at the moment of our call, and (2) the total length $T$ of the conversations of the $k$ others waiting, when our call occurred. We thus have

$$\gamma = \tau + T, \quad M_k\, e^{i\gamma\xi} = M_k\, (e^{i\tau\xi}\, e^{iT\xi}).$$

But clearly for a *given* $k$ the variables $\tau$ and $T$ are mutually independent. Thus it follows that

$$\varphi(\xi) = 1 - \alpha + \sum_{k=0}^{\infty} P(k) M_k\, e^{i\tau\xi}\, M_k\, e^{iT\xi}.$$

The expression $M_k\, e^{iT\xi}$ is easily simplified. Indeed the variable $T$ is the sum of $k$ mutually independent random variables, distributed according to the law $F(x)$ with the characteristic function $\psi(\xi)$. Thus

$$M_k\, e^{iT\xi} = [\psi(\xi)]^k \qquad (k = 0,1,2, \ldots)$$

and we find

$$\varphi(\xi) = 1 - \alpha + \sum_{k=0}^{\infty} [\psi(\xi)]^k P(k) M_k\, e^{i\tau\xi}.$$

We shall denote by $Q_k(x)$ the distribution function of the variable $\tau$ (the probability of the inequality $\tau > x$) for $k$ others waiting. Then

$$M_k\, e^{i\tau\xi} = -\int_0^\infty e^{i\xi x}\, dQ_k(x)$$

$$P(k)M_k\, e^{i\tau\xi} = -\int_0^\infty e^{i\xi x}\, P(k)dQ_k(x).$$

But the expression

$$P(k)[Q_k(x) - Q_k(x+dx)]$$

is the probability that our call will find the line occupied with $k$ calls waiting and that when this is so $x < \tau < x + dx$. This probability by virtue of lemma 4 equals

$$-\lambda dx \int_x^\infty u_k(t-x)dF(t),$$

as $dx \to 0$ to a higher degree of accuracy. Thus we get

$$P(k)M_k\, e^{i\xi\tau} = -\lambda \int_0^\infty e^{i\xi x}\, dx \int_x^\infty u_k(t-x)dF(t)$$

and consequently

$$\varphi(\xi) = 1 - \alpha - \lambda \int_0^\infty e^{i\xi x}\, dx \int_x^\infty \left\{ \sum_{k=0}^\infty [\psi(\xi)]^k\, u_k(t-x) \right\} dF(t),$$

from which by virtue of lemma 6

$$\varphi(\xi) = 1 - \alpha - \lambda \int_0^\infty e^{i\xi x}\, dx \int_x^\infty \exp\,[\lambda(t-x)\{\psi(\xi) - 1\}] \times$$

$$\sum_{k=0}^\infty \pi_k[\psi(\xi)]^k\, dF(t)$$

$$= 1 - \alpha - \lambda\chi(\xi) \int_0^\infty e^{i\xi x}\, dx \int_x^\infty \exp\,[\lambda(t-x)\{\psi(\xi) - 1\}]\, dF(t),$$

where it is assumed, as in § 39, that $\sum_{k=0}^\infty \pi_k[\psi(\xi)]^k = \chi(\xi)$. If, as in § 39, it is assumed that

$$\frac{\lambda}{i}\,[\psi(\xi) - 1] = \eta = \eta(\xi),$$

then it follows that

$$\varphi(\xi) = 1 - \alpha - \lambda\chi(\xi) \int_0^\infty e^{i(\xi-\eta)x}\, dx \int_x^\infty e^{i\eta t}\, dF(t)$$

$$= 1 - \alpha - \lambda\chi(\xi) \int_0^\infty e^{i\eta t}\, dF(t) \int_0^t e^{i(\xi-\eta)x}\, dx$$

$$= 1 - \alpha - \lambda\chi(\xi) \int_0^\infty e^{i\eta t}\, dF(t) \frac{e^{i(\xi-\eta)\,t} - 1}{i(\xi-\eta)}$$

$$= 1 - \alpha - \frac{\lambda\chi(\xi)}{i(\xi-\eta)} \int_0^\infty [e^{i\xi t} - e^{i\eta t}]\, dF(t)$$

$$= 1 - \alpha + \frac{\lambda}{i}\, \chi(\xi)\, \frac{\psi(\xi) - \psi(\eta)}{\xi - \eta}.$$

Finally, using the value of $\chi(\xi)$ given by lemma 7, we get

$$\varphi(\xi) = 1 - \alpha + (1-\alpha)\frac{\lambda}{i}\, \frac{\psi(\xi) - 1}{\xi-\eta}$$

$$= (1-\alpha) \left\{ 1 + \frac{\lambda}{i}\, \frac{\psi(\xi) - 1}{\xi - \frac{\lambda}{i}\, [\psi(\xi) - 1]} \right\}$$

$$= (1-\alpha) \frac{\frac{i\xi}{\lambda}}{\frac{i\xi}{\lambda} - [\psi(\xi) - 1]} = \frac{1 - \alpha}{1 - \alpha\, \frac{\psi(\xi) - 1}{is\xi}}.$$

The formula obtained in this fashion

$$\varphi(\xi) = \frac{1 - \alpha}{1 - \alpha\, \frac{\psi(\xi) - 1}{is\xi}} \tag{40.1}$$

can be regarded as the full solution of the problem in hand since the characteristic function $\varphi(\xi)$ of the required distribution function of the time of waiting has been expressed in terms of the given constants $s$ and $\alpha = \lambda s$ and the characteristic function $\psi(\xi)$ of the given distribution function $F(t)$ of the lengths of conversations.

As a first illustration of formula (40.1), the mean value $\bar{\gamma}$ of the time of waiting will be determined. Since $\varphi(0) = 1$,

$$i\bar{\gamma} = \varphi'(0) = \frac{\varphi'(0)}{\varphi(0)},$$

but since by virtue of (40.1)

$$\varphi(\xi) = \frac{(1-\alpha)is\xi}{is\xi - \alpha\,[\psi(\xi) - 1]},$$

it follows that

$$\frac{\varphi'(\xi)}{\varphi(\xi)} = \frac{d\log_e \varphi(\xi)}{d\xi} = \frac{1}{\xi} - \frac{is - \alpha\,\psi'(\xi)}{is\xi - \alpha\,[\psi(\xi) - 1]}$$

$$= \frac{\alpha\,\{\xi\psi'(\xi) - [\psi(\xi) - 1]\}}{\xi\,\{is\xi - \alpha\,[\psi(\xi) - 1]\}}.$$

The limit of this expression as $\xi \to 0$ can be found by L'Hôpital's rule and is equal to

$$\frac{\alpha\,\psi''(0)}{2(1-\alpha)is}.$$

The quantity

$$\psi''(0) = \int_0^\infty t^2\,dF(t) = -s_2$$

is the mathematical expectation of the square of the length of the conversation taken with a negative sign. We then find

$$i\bar{\gamma} = -\frac{\alpha}{2i(1-\alpha)}\,\frac{s_2}{s},$$

or

$$\bar{\gamma} = \frac{\alpha}{2(1-\alpha)}\,\frac{s_2}{s}.$$

This simple formula shows amongst other things that for a given loading of the line $\alpha$ and for a given mean length of the conversation $s$, the waiting time will, on the average, decrease with the dispersion of the lengths of conversations, i.e. the more standard these lengths are, the shorter the waiting time.

As a second illustration, consider the simple case of an exponential distribution

$$F(t) = e^{-\beta t}$$

of the lengths of conversations. In this instance,

$$s = \frac{1}{\beta}, \quad \psi(\xi) = \frac{\beta}{\beta - i\xi} = \frac{1}{1 - i\xi s},$$

from which

$$\frac{\psi(\xi) - 1}{is\xi} = \frac{1}{1 - i\xi s} = \psi(\xi),$$

and from (40.1)

$$\varphi(\xi) = (1-\alpha)\,\frac{1 - i\xi s}{1 - \alpha - i\xi s}$$

$$= 1 - \alpha + \alpha\,\frac{1-\alpha}{1 - \alpha - i\xi s}. \qquad (40.2)$$

But

$$\frac{1-a}{1-a-i\xi s}=\frac{\dfrac{1-a}{s}}{\dfrac{1-a}{s}-i\xi}$$

is the characteristic function of the exponential law of distribution with parameter $(1-a)/s$. Thus (40.2) shows that the distribution function of the waiting time is

$$P\{\gamma \geqslant t\}=(1-a)E(t)+a\exp\left[-\frac{1-a}{s}t\right] \qquad (t \geqslant 0)$$

where

$$E(t)=\begin{cases}1 & (t \leqslant 0)\\ 0 & (t > 0)\end{cases}.$$

This agrees with the formula

$$P\{\gamma > t\}=\pi e^{-(n\beta-\lambda)t},$$

which was obtained at the end of Chapter 9 in the case of $n$ lines. In fact, when $n = 1$, we have $\pi = a = \lambda s$, $n\beta - \lambda = \dfrac{1}{s} - \lambda = \dfrac{1-a}{s}$. Further, the term $(1-a)E(t)$ is equal to zero when $t > 0$, and for $t = 0$ becomes $1 - a = P\{\gamma = 0\}$.

# SECTION NOTES AND REFERENCES

(Numbers in square brackets refer to the bibliographical list following the present list of references)

§ 1 This elementary information is to be found in almost all present-day courses on the theory of probability; cf. in particular Gnedenko [1], Feller [3], Fry [4], Khintchine [5], Erlang [7].

§ 2 Apparently an original publication.

§ 3, 4 cf. § 1.

§ 5 cf. Fry [4].

§ 6 In this form, an original publication.

§ 7 Original publication.

§ 8 cf. Redheffer [9] where the same problem is solved in a different way.

§ 9, 10 For function $\varphi_0(t)$ cf. Palm [8]. For function $\varphi_k(t)$ $(k > 0)$ an original publication.

§ 11, 12 Original publication.

§ 13 The concept of a stream with limited after-effect comes from Palm [8]. The remaining contents of the section are original.

§ 14 cf. Palm [8].

§ 15, 16 Original publication.

§ 17, 18, 19, 20 cf. Feller [3], Fry [4], Erlang [7].

§ 21 This ergodic theorem is a particular instance of well-known general results of ergodic theory. It is here published in this form apparently for the first time.

§ 22, 23 Erlang [7].

§ 24, 25 Original publication. Particular instances of Palm [8].

§ 26 cf. Palm [8].

§ 27 Original publication.

§ 28, 29, 30, 31, 32, 33 cf. Palm [8].

§ 34, 35 cf. Kolmogoroff [2], Feller [3], Fry [4], Erlang [7].

§ 36, 37 cf. Erlang [7].

§ 38, 39, 40 All of Chapter 11 is a substantial rearrangement of the Author's article [6].

## BIBLIOGRAPHY(*)

[1] Gnedenko, B. V. (1958), *The theory of probability* (transl. of the 2nd Russian edn), Chelsea Publishing Co.

[2] Kolmogoroff, A. N. (1931), Sur le problème d'attente, *Mat. Sbornik*, **38**, No. 1–2, 101–106.

[3] Feller, W. (1957), *An introduction to probability theory and its applications* (Vol. 1, 2nd edn), Wiley.

[4] Fry, T. C. (1928), *Probability and its engineering uses*. Macmillan–Van Nostrand.

[5] Khintchine, A. Y. (1933), Asymptotische Gesetze der Wahrscheinlichkeitsrechnung, *Ergebn. Math.*, **2**, No. 4.

[6] Khintchine, A. Y. (1932), Mathematisches über die Erwartung vor einem öffentlichen Schalter, *Mat. Sbornik*, **39**, No. 4, 73–84.

[7] Copenhagen Telephone Co. (1948), *The life and works of A. K. Erlang*.

[8] Palm, C. (1943), Intensitätsschwankungen im Fernsprechverkehr, *Ericsson Technics*, **44**, 1–189.

[9] Redheffer, R. M. (1953), A note on the Poisson law, *Math. Mag.*, **26**, No. 2, 185–188.

(*)See Additional Note 5 on page 124.

# ADDITIONAL NOTES

by ERIC WOLMAN

Bell Telephone Laboratories
Holmdel, New Jersey, U.S.A.

These notes are intended to assist the reader in the few places where Khintchine's proofs have gaps or minor errors. In no case is the main line of argument seriously affected.

*Note* 1 (page 40). As Kendall has observed(*), the assertion of differentiability here is not justified by the preceding argument, and is in fact not always true.

The trouble originates in Section 9. Although the functions $H_k(\tau,t)$ used on page 38 are continuous in both variables, the limits $\Phi_k(t)$ in (9.2) need not be continuous in $t$. (It is easy to see how this comes about by sketching the $H_k$ for the deterministic process with $\lambda = 1$. In this case, each $\Phi_k$ is discontinuous at $t = k + 1$, where it is continuous on the right.) The functions $\varphi_k$ are then only *right*-continuous, which makes intuitive sense if the "period of length $t$" used in interpreting them (top of page 39) is an interval closed on the right. This background clarifies the argument of Section 10, which formally — since in the proof $\tau > 0$ — demonstrates the existence of *right-hand* derivatives $v_k'$ in (10.4). Thus there can be a set of measure zero on which the derivative $v_k'$ does not exist; but each right-hand derivative exists, and the integral formulae (10.8) are correct.

*Note* 2 (page 48). This part of the argument is valid only if both members (sides) of the last equation on page 47 are non-negative. Since the term $[-\psi_{r+1}(a)\,F_{r+2}(t)]$ in the right-hand member of that equation is at most zero,

$$\psi_r(a)[1 - \varphi_0(t) - F_{r+2}(t)] \leqslant \psi_{r+1}(a),$$

(*) KENDALL, D. G., *J. Internat. Statist. Inst.*, 1961, **29**, 2.

from which the fourth line on page 48 follows if the left-hand member is non-negative. But if that quantity is negative, we may argue as follows:

$$\psi_r(a)|1 - \varphi_0(t) - F_{r+2}(t)| = - \psi_r(a)[1 - \varphi_0(t) - F_{r+2}(t)]$$
$$= \psi_{r+1}(a)\, F_{r+2}(t) - \sum_{k>r} v_k(a)\, F_{k+2}(t)$$
$$\leqslant \psi_{r+1}(a)\, F_{r+2}(t) \leqslant \psi_{r+1}(a),$$

which was to be proved.

*Note* 3 (page 108). The two previous equations, and the text that follows them, should be ignored. They are unnecessary, and the second of these equations has not been established at this point in the argument. The text is correct if (39.2) comes immediately after the line "it follows that (39.1) is equivalent to the relation".

*Note* 4 (page 108). The argument of this paragraph is incorrect. The restriction imposed in the first sentence, that $T < u$, makes the final equation on page 108 a consequence of this argument only if $T < a$. This fact and the discussion of condition (1) (in line 5 of this paragraph) are appropriate to the limit as $T \to 0$; but (39.2) shows that the limit as $T \to \infty$ is required.

Since, in the system under study, calls can be delayed, we should define a quantity $L_T^{**}(a,b)$, the total length of potential conversations of length $(a,b)$ corresponding to calls entering the system during the segment $(0,T)$. Of course $L_T^{**}(a,b)$ can differ from $L_T(a,b)$ if calls are waiting at time 0 or at time $T$. Indeed, if $m$ calls are waiting at 0 and $n$ are waiting at $T$, then

$$L_T^{**}(a,b) - L_T(a,b) = (n - m)\zeta$$

where $a \leqslant \zeta \leqslant b$. Therefore

$$|ML_T^{**}(a,b) - ML_T(a,b)| = \zeta|M(n - m)|. \qquad \text{(N4.1)}$$

Inherent in all of Chapter 11 is the assumption of stationarity. The random variable $\gamma$ is assumed to have a distribution function (*vide* (40.1)), and the distribution $\pi_k$ introduced in Section 38 makes no sense without stationarity. The existence of a proper distribution function for

$\gamma$ is equivalent in this context, by a theorem of Lindley[*], to the condition $\alpha < 1$; and this condition is implied in Khintchine's definition of $\alpha$. But in order to prove lemma 1, we must put (39.2) in the form

$$\lim_{T\to\infty} \frac{1}{T} ML_T^{**}(a,b) = -\lambda \int_a^b u\, dF(u). \tag{N4.2}$$

This follows from (N4.1) if $M(n)$ and $M(m)$ are finite; in which case, by stationarity, they are equal. The finiteness of $M(m)$ is a much stronger condition than is the existence of the $\pi_k$. It is equivalent to the finiteness of $\bar{\gamma}$ ($= M(\gamma)$), and this, as we see on page 117, requires the existence of $s_2$, the second moment of $F$. (This requirement is related directly to the finiteness of $M(m)$ in Kendall's paper [†] of 1951.)

With $s_2$ assumed finite, so that only (N4.2) must be verified, the proof proceeds as in the text. The expected number of calls entering the system during $(0,T)$ is *exactly* $\lambda T$, of which a fraction $-\, dF(u)$ will have lengths of conversations $(u, u + du)$. Therefore

$$ML_T^{**}(u, u + du) = -\lambda T(u + \theta du)\, dF(u),$$

so that

$$ML_T^{**}(a,b) = -\lambda T \int_a^b u\, dF(u).$$

From this follows (N4.2), and lemma 1 also.

The essential element of this corrected proof is its dependence upon $L_T^{**}$, whose mean value we know explicitly, rather than upon $L_T$.

An interesting aspect of Chapter 11 is that its general approach to the problem of finding the distribution of $\gamma$, as manifested in the corrected proof of lemma 1, seems to require that we assume for $F$, the distribution function of the length of conversations, not only a finite mean (*vide* page 106, paragraph 2) but also a finite second moment. However,

---

[*] LINDLEY, D. V., The Theory of Queues with a Single Server, *Proc. Camb. Phil. Soc.*, 1952, **48**, 277–289; *vide* page 281.

[†] KENDALL, D. G., Some Problems in the Theory of Queues, *J. Roy. Statist. Soc.* (*B*), 1951, **13**, 151–185.

the Pollaczek–Khintchine formula itself ((40.1) on page 116) is valid without this restriction and holds whenever $F$ has a finite mean and $a < 1$. (See, for example, pages 68–9 of the book by Takács(*).)

*Note* 5 (page 120). In addition to the references given in this bibliography, the reader may wish to consult two papers which complement and complete the theory of streams of events described in Part I of this book. These papers, both by Khintchine, are:

Sequences of Chance Events without After-Effects, *Theory of Probability and its Applications*, 1956, **1**, 1–15;

On Poisson Sequences of Chance Events, *Theory of Probability and its Applications*, 1956, **1**, 291–297.

(Page numbers refer to the English version of this journal, a translation from the Russian, published by the Society for Industrial and Applied Mathematics.)



(*) TAKÁCS, LAJOS, *Introduction to the Theory of Queues*, Oxford University Press (New York). 1962.